Cockpit Weather Decisions

Other books in the Practical Flying Series

Light Airplane Navigation Essentials
Paul A. Craig

Handling In-Flight Emergencies
Jerry A. Eichenberger

Cockpit Resource Management: The Private Pilot's Guide
Thomas P. Turner

The Pilot's Guide to Weather Reports, Forecasts, and Flight Planning—2nd Edition
Terry L. Lankford

Weather Patterns and Phenomena: A Pilot's Guide
Thomas P. Turner

Cross-Country Flying
Jerry A. Eichenberger

Avoiding Mid-Air Collisions
Shari Stamford Krause, Ph.D.

Flying in Adverse Conditions
R. Randall Padfield

Advanced Aircraft Systems
David Lombardo

Understanding Aeronautical Charts—2nd Edition
Terry T. Lankford

Aviator's Guide to Navigation—3rd Edition
Donald J. Clausing

Learning to Fly Helicopters
R. Randall Padfield

ABCs of Safe Flying—3rd Edition
David Frazier

Flying VFR in Marginal Weather—3rd Edition
R. Randall Padfield

The Aviator's Guide to Navigation—3rd Edition
Donald J. Clausing

Better Takeoffs & Landings
Michael Charles Love

The Art of Instrument Flying—3rd Edition
J. R. Williams

Aviator's Guide to GPS—2nd Edition
Bill Clarke

Cockpit Weather Decisions

Terry T. Lankford

McGraw-Hill
New York San Francisco Washington, D.C. Auckland Bogotá
Caracas Lisbon London Madrid Mexico City Milan
Montreal New Delhi San Juan Singapore
Sydney Tokyo Toronto

Library of Congress Cataloging-in-Publication Data

Lankford, Terry T.
Cockpit weather decisions / Terry T. Lankford.
p. cm. — (Practical flying series)
Includes bibliographical references and index.
ISBN 0-07-036719-1
1. Meteorology in aeronautics. 2. Airplanes—Piloting.
I. Title. II Series.
TL556.L357 1997
629.132'4—dc21 97-24639
CIP

McGraw-Hill

A Division of The ***McGraw-Hill*** *Companies*

1 2 3 4 5 6 7 8 9 0 FGR/FGR 9 0 2 1 0 9 8 7

ISBN 0-07-036719-1

The sponsoring editor for this book was Shelley Carr, the editing supervisor was Scott Amerman, and the production supervisor was Pamela Pelton. It was set in Times by Tanya Howden and Kim Sheran of McGraw-Hill's Professional Group Composition Unit, in Hightstown, NJ.

Printed and bound by Quebecor/Fairfield.

McGraw-Hill books are available at special quantity discounts to use as premiums and sales promotions, or for use in corporate training programs. For more information, please write to the Director of Special Sales, McGraw-Hill, 11 West 19th Street, New York, NY 10011. Or contact your local bookstore.

This book is printed on recycled, acid-free paper containing a minimum of 50 percent recycled, de-inked fiber.

PFS

Contents

Preface ix

Introduction xi

1 The Realm of Flight 1
Structure of the Atmosphere *2*
  The Troposphere *2*
  The Tropopause *3*
  The Stratosphere *3*
  The Jet Stream *3*
Atmospheric Properties *5*
  Temperature *5*
  Pressure *8*
  Moisture *9*
  Density *11*
Altimetry, Airspeed, and Aircraft Performance *13*
  The Altimeter *14*
  Airspeed *17*
  Aircraft Performance *17*
Strategies *18*
Updating Weather En Route *21*

2 Ever-Changing Weather 23

Lapse Rate *24*
- Stability *28*

Vertical Motion in the Atmosphere *31*
Low Ceilings and Visibility *33*
- Fog *37*
- Haze and Smoke *41*
- Dust, Sand, and Volcanic Ash *43*
- Precipitation *45*
- Terrain Obscurement *47*

Strategies *48*

3 Motion of the Air 53

General Circulation *54*
Winds *58*
- Sea and Land Breezes *60*
- Mountain and Valley Winds *60*
- Downslope Winds *61*
- Mountain Waves *61*

Turbulence *65*
- Mechanical Turbulence *66*
- Thermal Turbulence *67*
- Windshear-Induced Turbulence *68*
- Frontal Turbulence *70*
- Clear Air Turbulence *70*
- Wake Turbulence *71*

Strategies *72*

4 Signpost in the Sky 79

Cloud Types *80*
- Low Clouds *81*
- Middle Clouds *83*
- High Clouds *85*
- Clouds with Vertical Development *88*

Precipitation *92*
Icing *95*
- Rime Ice *97*
- Clear Ice *97*
- Rime Ice and Clear Ice (Mixed Ice) *97*
- Structural Icing *99*
- Induction or Carburetor Icing *100*

Strategies *103*

5 Families of the Air 109
Air Masses *110*
Low Pressure Areas and Troughs *113*
High Pressure Areas and Ridges *114*
Thunderstorms *114*
Limited State Thunderstorms *118*
Steady State Thunderstorms *120*
Thunderstorm Hazards *122*
Convective Low-Level Windshear *128*
Windshear Recognition *129*
Takeoff, Approach, and Landing Precautions *132*
Windshear Recovery Technique *132*
Strategies *133*

6 Air Masses in Conflict 139
The Life Cycle of a Front *141*
Cold Fronts *143*
Warm Fronts *149*
Occluded Fronts *152*
Stationary Fronts *154*
Strategies *155*

7 Where's the Front? 161
Upper-Level Troughs and Ridges *163*
Upper-Level Lows and Highs *168*
The Jet Stream Revisited *172*
Vorticity *173*
Convergence and Divergence *176*
Hurricanes *178*
Putting It Together *181*
Strategies *183*

Glossary 187

Appendix A: Charts and Graphs 195

Appendix B: METAR/TAF Codes 199

Index 211

Preface

I obtained a private pilot certificate in 1967 through an Air Force aero club in England. Back in the United States, under the G.I. Bill, I obtained commercial and flight instructor certificates, along with an instrument rating. Subsequently, as a full-time flight instructor, I earned a Gold Seal, and I recently added multiengine to my instructor certificate. During this period, I owned and operated a Federal Aviation Administration (FAA) approved ground school in the Los Angeles area.

I have owned two airplanes, both Cessna 150s, that have taken me across the country twice. I have flown in Hawaii, Canada, and Mexico, as well as Europe. I remain an active pilot and flight instructor, having flow three trips from the San Francisco Bay area to Oshkosh, Wisconsin, in the last 5 years.

A commercial pilot certificate and instrument rating qualified me for my present position with the FAA as an Air Traffic Control Specialist (Station). I've been a Flight Service Station (FSS) specialist for more than 20 years, bringing together two interests: aviation and weather.

Because of my position with the FAA, I have been privileged to attend three weather courses at the FAA Academy—each equivalent to a college semester. The first was in 1974 as part of the FSS basic course. In 1978 I attended the Pilot Weather Briefer Refresher Course, which qualified me to teach the class in the field. In 1986 I completed the En Route Flight Advisory Service (Flight Watch) course.

Over the years I've found no comprehensive source of weather information for pilots. Most weather texts have been dry and complex, and they often didn't relate to the real world. It is exasperating to read an article or book about aviation weather that espouses the need to understand meteorological terms but fails to include a definition or explain the application of the concept.

I have also discovered that almost all aviation weather texts, both government and commercial, suffer from poor organization and omission of required subjects. Most contain too much detail in climatology and general meteorological theory that does not relate to flight operations.

As a pilot, I want to know how the weather affects me and my aircraft during flight—the bottom line. This book is written for the pilot, from the pilot's point of view. It does not require the memorization of countless formulas, the use of calculus, or a degree in meteorology. Many complex phenomena can be simplified while explanations are still given about their operational impact on aviation. (This simplification, undoubtedly, will cause some consternation in the meteorological community, but, our goal is operational, not theoretical, knowledge.)

Publications that attempt to condense weather theory, reports, forecasts, and charts into a single publication cannot adequately cover the subject. Weather is too complex and extensive for a single source. Weather reports, forecasts, and chart formats change relatively rapidly, requiring frequent revisions. The FAA discovered this many years ago and established its publications *Aviation Weather* AC 00-6 and *Aviation Weather Services* AC 00-45. Weather reports and forecasts are not within the scope of this book, although we do direct the reader to the reports and forecasts where related information can be found. A detailed discussion of weather reports and forecasts—decoding, interpretation, application, and other flight-planning topics—are presented in McGraw-Hill's companion publication, *The Pilot's Guide to Weather Reports, Forecasts, and Flight Planning,* Second Edition.

My goal is take all the information available, sift out the salient material from a pilot's viewpoint, and present it in an organized and entertaining way that applies to flight operations.

A majority of the photographs, including satellite images, were taken in the western United States. There are numerous references to weather phenomena in the West, due to the fact that that's where I live. I have greater access to material in this part of the country. However, the material in the book applies to the entire United States, Canada, and Mexico, and, in fact, to all midlatitude areas of the earth.

I am greatly indebted to many people for their generous help, guidance, and advice. They are too numerous to be listed in full. Among them are the meteorologists of the National Weather Service (NWS) at the Aviation Weather Center in Kansas City, the FAA Academy in Oklahoma City, and local and regional NWS offices, plus light service station specialists that I have been privileged to know, especially at the Oakland, California, FSS. A special thanks goes to the members of the National Weather Association's Aviation Meteorology Committee, whose help is greatly appreciated. Finally, I must acknowledge the pilots who have allowed me to assist them and in turn provided me with the best weather education possible. This book is dedicated to these people.

Introduction

Weather affects a pilot's flying activity more than any other physical factor. Most pilots agree that weather is the most difficult and least understood subject in the training curriculum. Surveys indicate that many pilots are uneasy with, or even intimidated by, weather. In spite of these facts, or perhaps because of them, weather training for pilots consists of the bare bones, only enough to pass the written test, while weather-related fatal accident rates remain relatively unchanged. These observations were reaffirmed in the 1992 National Aviation Weather Program Plan. The plan identified a number of unmet needs. Among these was the necessity to improve aviation weather education for pilots and weather briefers.

The Aircraft Owners and Pilots Association's (AOPA) Air Safety Foundation publication *Safety Review General Aviation Weather Accidents* (*An Analysis & Preventive Strategies*) *1996* provides a detailed analysis of general aviation accidents. We recommend this publication, which should be part of every pilot's library.

With respect to accident statistics and scenarios, some may say, why emphasize the negative? Our goal, along with AOPA's Air Safety Foundation and the Federal Aviation Administration, is accident prevention. People learn either through training or experience. Unfortunately, the definition of learning by experience is when "the test comes before the lesson." We hope, through the incidents and scenarios presented in this book, to help prevent pilots from becoming accident statistics.

While we're on the subject of accident scenarios, let's concede that hindsight is always 20/20. It's easy to sit back in a favorite armchair and analyze and criticize someone else's performance and decisions. When we review an accident in this book, or any other publication or forum for that matter, let's not judge or attempt to blame. Our goal is prevention. Therefore, if the reader perceives any judgment or blame in any of the incidents or scenarios in this book, it is strictly unintentional; that is not our purpose.

INTRODUCTION

Periodically, we talk about minimums. Minimums are just that—minimums. Minimums do not necessarily mean safe. Each pilot, whether rated for visual flight rules (VFR), or instrument flight rules (IFR), needs to set his or her own safe minimums. For example, it makes no sense to depart VFR with an overcast of 500 feet, visibility 1 mile in a mountainous area. It may be technically legal, but it's not safe! Nor does it make any sense to fly in an area of thunderstorms in an aircraft without storm-detection equipment. Again, it is legal, but it's not safe! Minimums and the decision to initiate a flight should be based on our own training and experience and the capabilities of the aircraft and equipment we fly. In subsequent chapters, we explore how a number of accidents occurred even though the pilot had "legal minimums."

During the past 10 years, approximately 27% of all general aviation accidents involved adverse weather conditions. Of these, 30% involved fatalities. The biggest causal factor, as in the past, involved continued VFR flight into instrument meteorological conditions (IMC). This factor alone accounts for 56% of fatalities for all weather-related accidents. By a wide margin, the most dangerous situation is VFR flight into IMC. Therefore, a major theme of this book is risk management. That is, how can we as pilots avoid becoming accident statistics?

An essential part of flight preparation concerns the weather. No matter how short or simple the mission, Federal Aviation Regulations (FAR) place the responsibility for flight preparation on the pilot. To effectively use all available sources, a pilot must have a basic but sound understanding of weather phenomena and theory.

This book provides information that not only helps to pass written and practical examinations but also prepares pilots—whether student, airline transport, recreational, or biz-jet—to understand the environment within which they operate. Pilots will be able to recognize atmospheric conditions conducive to icing, turbulence, thunderstorms, and other weather hazards.

Various government and nongovernment publications—books, articles, pamphlets, and videos—have attempted to provide and improve weather education for pilots. Most books do not relate weather theory and phenomena to actual flight situations. Articles and pamphlets typically can only address one issue. The same may be said about videos. The AOPA Air Safety Foundation produced an informative and entertaining *Tactical Weather Workshop* program. Flight scenarios were based on a turbo, pressurized, radar- and stormscope-equipped Cessna 210. Unfortunately, many of us don't have the opportunity to fly this type of equipment. Flying sophisticated, well-equipped aircraft allows the pilot many more options when dealing with the weather. While most of the scenarios used in this book relate to nonturbo, single-engine aircraft, the principles apply to all.

Technical meteorological concepts and terms are translated into language anyone can easily understand. On the other hand, such subjects as vorticity, microbursts, and upper-level weather systems are not omitted because of their complexity. Discussions include applying weather to VFR as well as IFR flights and high-level as well as low-level operations.

Rather than splitting out aviation weather hazards, we've included these phenomena in appropriate chapters, eliminating repeated material and the need to refer back to previous sections. We've tried to organize subjects using the building-block approach. Information from one section is used to increase understanding of subsequent sections. Elements of one chapter are used as a foundation for the material in later chapters. In this way the reader is not overwhelmed by any section or chapter. Unfortunately, we occasionally run into the "chicken or the egg" syndrome. We may have to introduce a topic, then provide details in subsequent sections. Because of this, the reader should not be overly concerned if any one element or concept is not completely clear on initial reading.

Throughout the book we indicate where information on specific meteorological phenomena can be located. For example, when we discuss frontal systems, we indicate that these weather patterns can be located on National Weather Service (NWS) generated surface analysis charts. Charts produced by private vendors typically provide less data than NWS products. Therefore, readers may not find all the data we mention on non-NWS charts. However, like weather reports and forecasts, the charts themselves are not within the scope of this text.

Each chapter contains a section titled "Strategies." These sections describe specifically how to handle weather hazards. Unfortunately, it's difficult to teach judgment—the ability to evaluate facts and come to a rational, safe decision. However, I do believe that judgment, like common sense, is largely built on training and experience. My goal is to share my training and experience. In this way we can, hopefully, avoid learning only by experience.

Chapter 1 begins with a description of the atmosphere. Since aircraft (we're not going to count the space shuttle) only fly within the lower two layers of the atmosphere, the *troposphere* and *stratosphere,* only these layers are discussed, along with their boundary, the *tropopause.* A detailed discussion of the jet stream describes this phenomena. Since the influence of the jet stream on weather—both surface and aloft—has such significance, this subject is visited in subsequent chapters.

The chapter also discusses atmospheric properties that affect aviation weather, including temperature, pressure, moisture, and density. Without moisture we would not have weather as we know it. Along with moisture, atmospheric density is the key to understanding most weather phenomena; its importance cannot be overemphasized.

With an understanding of atmospheric properties, we move on to discussions on altimetry, airspeed, and aircraft performance. These discussions include how the atmosphere affects aircraft performance during all phases of flight. Next, we present strategies to deal with high temperatures and conditions that produce high-density altitude. The final section contains an overview of sources of information that allow a pilot to make intelligent inflight decisions.

Chapter 2 discusses the effects of stability, or lack therefore, in the atmosphere. This chapter contains a detailed section on *lapse rate.* Lapse rate is essential to an understanding of vertical motion, and vertical motion is a primary cause of weather. Included are the concepts of the level of free convection and lifted condensation level. Both are important to an understanding of cloud development and thunderstorms.

Next we discuss low ceilings and visibilities and their effects on aviation operations. This section describes fog, haze and smoke, dust, sand, and precipitation. Over the last decade, another aviation weather hazard has been explored—volcanic ash. A discussion of volcanic ash is included in this chapter. Terrain obscurement constitutes an aviation weather hazard of itself. Similar to IFR conditions, terrain obscurement typically applies to mountainous regions. The chapter includes strategies to deal with the effect of low ceilings and visibilities, both for IFR as well as VFR operations.

With an understanding of the atmosphere, its properties, and stability, we move on in Chapter 3 to pressure patterns and wind. The wind is a mover of atmospheric properties from one location to another. These properties are modified by the terrain over which they pass. Wind and terrain are additional mechanisms that produce and modify weather. This chapter examines local winds, with an emphasis on their effect on aviation operations.

A change of windspeed with height or horizontal distance causes shear. The effects of windshear on flight operations are discussed. Convective low-level windshear, with all its perilous implications, is presented in Chapter 5.

A detailed discussion of turbulence, its causes, location, and cloud types related to nonconvective turbulence are presented. The significance of rotor clouds, standing lenticular, and Kelvin-Helmholtz (K-H) windshear clouds is discussed. Because of its significant hazard, we included an artificial form of turbulence—wake turbulence. The chapter ends with strategies to minimize or avoid the effects of turbulence.

The effects and significance of clouds and precipitation are presented in Chapter 4. The effect of terrain on the weather is included. This chapter contains a detailed section on the effects of aircraft icing—induction, airframe, and carburetor. The final section provides strategies to minimize or avoid icing hazards.

We debated about including a cloud code chart. Since we are restricted to black and white photographs and significant cloud pictures are included in appropriate sections, we decided a separate cloud code chart would not be necessary. For those who would like an excellent, detailed, color cloud chart, "The Cloud Chart" is available through the National Weather Association, 6704 Wolke Court, Montgomery, Alabama, 36116-2134.

Chapter 5 discusses air masses. Air masses have specific characteristics that affect aviation. These characteristics are modified by underlying terrain. A knowledge of these phenomena can explain many weather conditions and hazards. For example, high pressure usually means good weather, but not always. Moisture trapped under high-pressure inversions may create widespread "zero-zero" conditions that can last for weeks. Strong pressure gradients within high-pressure areas can cause severe turbulence.

Thunderstorms contain just about every weather hazard, including the potential for high-density altitude. Thunderstorm development, along with its hazards, are presented in this chapter. Because of its significance, a discussion of microbursts, related phenomena, and recognition and recovery techniques is included. The final section, "Strategies," provides a comprehensive discussion of thunderstorm-avoidance procedures.

Frontal systems, their formation, and their dissipation have a major impact on flight operations. Chapter 6 covers these phenomena from birth, through evolution, to dissipation. Most texts cover only classical fronts—predominately those east of the Rockies. Because of mountainous terrain in the West, a frontal surface may have a depth of hundreds of miles. This chapter covers these phenomena throughout the United States and

North America. Like other chapters, the last section contains strategies for dealing with frontal weather.

The final chapter, Chapter 7, is devoted to nonfrontal weather systems. An all-too-often, common misconception is that the absence of a front means the absence of significant weather. This is not the case. Troughs, ridges, and vorticity are major weather producers. A pilot's weather education—even for those who fly below 10,000 feet—is woefully lacking without a sound knowledge of upper-level weather systems and their effect on surface weather. A front under a ridge aloft may provide no weather at all. On the other hand, a front supported by an upper trough and the jet stream produces the most severe weather anywhere. Finally, there is the hurricane, possibly the best example of a nonfrontal weather-producing system. We included a section called "Putting It Together." Here we combine the effects of our three-dimensional atmosphere. From this discussion we can see how all atmospheric phenomena comes together to produce our ever-changing weather. The last section, as in previous chapters, provides strategies for flying nonfrontal weather systems.

A glossary is included to serve as a quick reference for terms and concepts.

We have mentioned that weather reports and forecast are not within the scope of this book. However, with the implementation of the METAR and TAF codes on June 1, 1996, we decided to include Appendix A, Charts and Graphs, and Appendix B, METAR/TAF Codes, to cover this change and provide supplementary information. These appendixes do not attempt to interpret or apply the codes. There have been several changes since implementation and we have attempted to cover those changes and some proposed changes that may occur in the future.

Throughout the text we have used anecdotes to illustrate important points. Incidents are not intended to disparage or malign any individual, group, or organization; the sole purpose is illustration.

The following chapters, hopefully with a little humor, explain basic weather theory and phenomena that apply to aviation. Armed with this knowledge and a commitment not to push the weather, our aircraft, or ourselves, we can realistically realize a growing safety record as we enter the twenty-first century.

1
The realm of flight

THE EARTH IS UNIQUE IN THE SOLAR SYSTEM. EVOLUTION HAS provided a thin layer of gas covering the planet. The atmosphere consists of about 78% nitrogen, 21% oxygen, and lesser amounts of argon, carbon dioxide, and other gases—including water vapor, which is essential to our story. The types and amount of gases has only one implication to our weather discussion: If the gases were in a different ratio, life on earth as we know it would not exist.

Another unique, and essential, property of the earth is its temperature and pressure. The average temperature and pressure at the earth's surface allows water to exist as a liquid. If this were not true, we would not have weather. (We use the term *average* an awful lot. For our purposes, average is an arithmetic mean. Add up all the temperatures or pressure around the world and divide by the total number. Actual conditions on the earth or in the atmosphere are rarely average, as we shall see.)

The generation of weather requires an energy source. In the solar system, the sun provides virtually all the energy available to the earth's weather machine. Thus, we may attribute all our weather to the sun. So, we have a planet where water can exist as a liquid, with a constant, steady source of energy. The result is an atmosphere that sustains life and produces weather.

The earth's atmosphere has evolved over the past four and a half billion years. About 400 million years ago, the oxygen level became high enough for land animals to develop. Modern humans have been around for about 10,000 years. Aren't you glad you waited 'til now to study the weather?

In a subsequent section of this chapter, we apply our knowledge of the theory of the atmosphere to altimetry, airspeed, and aircraft performance. Atmospheric density can significantly affect all three. The following section contains strategies to avoid or reduce the effects of adverse atmospheric density.

Throughout the text, we emphasize the need to update weather information en route. After all, the *cockpit weather decision* depends on all available information. The final section discusses the sources and procedures for obtaining this information. So let's begin.

STRUCTURE OF THE ATMOSPHERE

The atmosphere can be divided into various layers or spheres, as illustrated in Fig. 1-1. Each sphere has unique properties. Based on temperature, the atmosphere is typically divided into the following:

- Troposphere
- Stratosphere
- Mesosphere
- Thermosphere
- Exosphere

Since most weather occurs within the troposphere, and most flying takes place in the troposphere and stratosphere, we concern ourselves with these two layers.

The Troposphere

Beginning at the earth's surface, the troposphere extends to an average altitude of about 7 miles. Thickness varies from the equator to the poles, as shown in Fig. 1-1. The troposphere extends higher over the equator than the poles and higher in summer than in winter. Why? Warm columns of air are taller than cool columns. Near the equator, the sun heats the surface, which in turn warms the air. Therefore, the troposphere extends to a higher altitude at the equator than at the poles. The same rationale explains why the troposphere extends to a higher altitude in summer; more solar radiation during the summer months makes the troposphere warmer.

The troposphere contains about three-quarters of the atmosphere by weight and almost all the water vapor. Nearly all clouds and weather occur within this layer.

The average thickness of the troposphere is 25,000 to 30,000 feet at the poles and between 55,000 and 65,000 feet at the equator. Average temperature decreases with altitude, from 15°C at sea level to about −57°C at 36,000 feet. (More about the average or *standard atmosphere* later in this chapter.) Pressure decreases with altitude from an average of 1013 millibars (mb) at sea level to about 230 mb at 36,000 feet. In the troposphere, windspeed tends to increase with height.

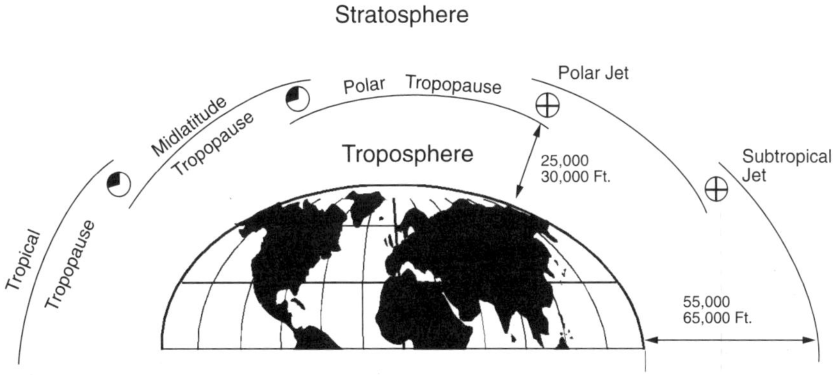

Fig. 1-1. *The atmosphere can be divided into various layers or spheres. Each has unique properties.*

The Tropopause

The boundary between the troposphere and the stratosphere is known as the tropopause. In this layer temperature remains relatively constant with altitude. Pressure continues to decrease with height.

As shown in Fig. 1-1, the tropopause is not a continuous layer. There are breaks in the tropopause. On the average, breaks occur between the polar tropopause and midlatitude tropopause and the midlatitude tropopause and the tropical tropopause. Where these breaks occur lies the jet stream. In Fig. 1-1, the bullets on the left represent the jet stream coming out of the illustration; the arrow feathers, on the right, represent the jet going into the illustration. In the northern hemisphere the jet stream generally blows from west to east. (More about the jet stream later in this chapter.)

The tropopause acts as a lid that resists the exchange of air between the troposphere and the stratosphere. It is why almost all water vapor and weather is found in the troposphere.

The Stratosphere

Above the tropopause is the stratosphere. The average altitude of this layer is from 7 miles to about 22 miles (115,000 feet). There is a slight increase in temperature with height. Pressure continues to decrease. Virtually no water vapor or clouds are found in this layer.

Ozone reaches its maximum concentration near the top of the stratosphere. Ozone is important because it absorbs most of the deadly ultraviolet rays from the sun. Above the stratosphere are the mesosphere, thermosphere, and exosphere respectively.

The Jet Stream

The jet stream was virtually unknown until World War II when pilots flying at high altitudes reported turbulence and tremendously strong winds. These winds blew from west to east near the top of the troposphere. Not until 1946 was the jet stream fully recognized as a meteorological phenomenon.

Sharp horizontal temperature differences cause strong pressure gradients that result in the jet stream. Refer to Fig. 1-2. Note the strong temperature gradients illustrated by *isotherms*—lines of equal temperature—(dashed lines) in Fig. 1-2. Temperature changes rapidly with height. The atmosphere compensates for extremely cold air in polar regions with relatively warmer air above. This relatively warmer air above the tropopause extends well up into the stratosphere. In such zones, the slope of constant pressure surfaces increases with height. Since fronts lie in zones of temperature contrast, the jet is closely linked with frontal boundaries. When windspeed becomes strong enough, the flow is termed a jet stream. *Isotachs*—lines of equal windspeed—increase to a maximum in the jet core.

A jet stream is a narrow, shallow, meandering area of strong winds embedded in breaks in the tropopause (Fig. 1-1 and Fig. 1-2). Two such breaks occur in the northern hemisphere: the polar jet, located around 30° to 60° north latitude at an approximate height of 30,000 feet, associated with the polar front, and the subtropical jet, around 20° to 30° north latitude at approximately 39,000 feet. To be classified a jet stream, winds must be 50 knots or greater; although, jet stream winds generally range between 100 and 150 knots, winds can reach 200 knots along the east coast of North America and Asia in winter when temperature contrasts are greatest.

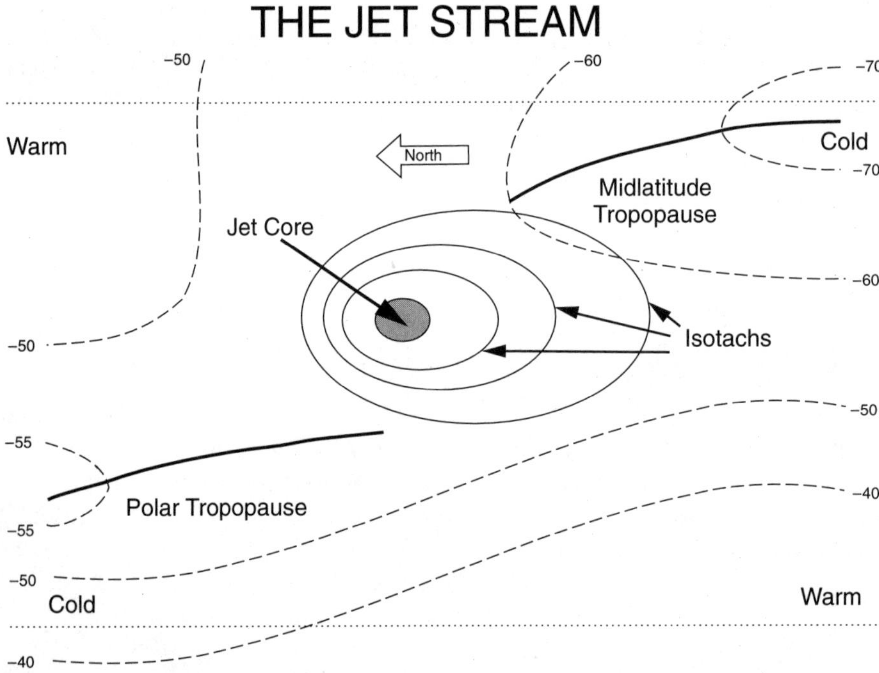

Fig. 1-2. *Sharp horizontal temperature differences cause strong pressure gradients that result in the jet stream.*

A "jet" is most frequently found in segments 1000 to 3000 miles long, 100 to 400 miles wide, and 3000 to 7000 feet deep. The strength of the jet stream increases in winter in mid and high latitudes when temperature contrasts are greatest and shift south with the seasonal migration of the polar front.

The presence of jet streams has significant operational impact on flight operations. The jet stream can cause a significant headwind component for westbound flights, increasing fuel consumption and requiring additional landings.

Another factor associated with the jet is windshear turbulence. Windshear is caused by a change in windspeed, either horizontal or vertical, or direction. A significant wind change over a relatively small distance can result in severe turbulence. Maximum jet stream turbulence tends to occur above the jet core and just below the core on the north side, as seen in Fig. 1-2. Notice that the maximum curvature (greatest change in direction) and closest spacing (greatest change in horizontal and vertical distance) of the isotachs surrounding the jet core occur on the north side of the jet. Additional areas of probable turbulence occur where the polar and subtropical jets merge or diverge. With an average depth of 3000 to 7000 feet, a change in altitude of a few thousand feet often takes the aircraft out of the worst turbulence and strongest winds.

ATMOSPHERIC PROPERTIES

For our discussion a *property* is a characteristic trait or peculiarity. The earth has weather because of its atmosphere, water, and the energy from the sun. The manifestations of weather result from these properties.

As aviators we're interested in those atmospheric properties that affect our flying activities—temperature, pressure, moisture, and density. Two of our most important primary flight instruments operate on these properties—the altimeter and the airspeed indicator. Therefore, it's appropriate to include discussions of these flight instruments. Additionally, atmospheric properties affect aircraft performance. The impact of the atmosphere on aircraft performance is included in this chapter.

Temperature

Temperature is a measure of the average speed of molecules. But let's back up for a moment. *Heat* is the total energy of the motion of molecules; heat is a form of energy, and it has the ability to do work. Temperature can, therefore, be related to the amount of energy available in the atmosphere. And energy drives the weather.

The amount of heat required to raise the temperature of air, or the amount of heat lost when cooled, is known as heat capacity. Energy is released or absorbed during the process. How this relates to aviation weather becomes apparent in subsequent chapters.

A concept we continually use is hot and cold, warm and cool. For our purposes, these terms are relative. For example, we may refer to a parcel of air, or an air mass, as warm. By this we mean a parcel of air that is warmer than the surrounding air, or an air mass warmer than the surface or warmer than an adjacent air mass. The parcel or air mass may have a temperature of $-10°$, but the surrounding air, surface, or adjacent air mass may have a temperature of $-15°$. Therefore, the parcel or air mass is warmer in relation to its surroundings.

It's important to understand how the sun heats the earth. Both dry air and water vapor are transparent to visible light, which constitutes most of the solar radiation coming from the sun. So what happens to the solar radiation other than visible light? A small amount is absorbed and heats the atmosphere. Some is reflected back to space by clouds and the atmosphere, and some is reflected by the earth's surface. The majority of the sun's energy (approximately 33%) is absorbed by the earth's surface. This radiant energy is converted to heat and raises the temperature of land and water surfaces.

During the day the earth's surface absorbs solar radiation and in turn the atmosphere near the surface is heated by conduction. *Conduction* is the process of heat transfer by contact from one substance to another. At night the earth radiates heat back into space. This process is known as *terrestrial radiation* or *radiational cooling*. This process, in turn, cools the atmosphere in contact with the surface.

The amount of energy absorbed by the earth depends on several factors. One factor is the angle at which solar radiation strikes the surface. More energy is absorbed at the equator than at the poles; more during summer than winter; more during midday than early morning or late afternoon. At night the surface radiates energy back into space, cooling the earth and the atmosphere at and near the surface.

Different surfaces absorb and reflect different amounts of the sun's radiation. Dense cloud layers greatly reduce the amount of energy reaching the surface. Snow-covered areas reflect most incoming solar radiation. Land areas absorb radiation to a much greater degree than water surfaces, with different land surfaces absorbing different amounts of radiation. For example, plowed fields absorb more radiation than green pastures, and green pastures absorb more than rivers or lakes. As a result, land becomes warmer during the day, while the temperature of water areas remains relatively constant. At night the land surfaces radiate heat back into the atmosphere, while water areas, again, remain relatively unchanged. Additionally, large bodies of water tend to retain heat much longer than land areas. This property is why temperatures tend to be more moderate along coast lines and why deep lakes do not freeze in the winter.

Liquid water droplets, ice particles, and water vapor are transparent to incoming solar radiation but opaque to outgoing terrestrial radiation. At night clouds absorb nearly all terrestrial radiation and reflect it back to the surface. This is why temperatures remain relatively cool on cloudy days and relatively warm during cloudy nights. In dry, desert areas, terrestrial radiation cools the land rapidly after sunset, but humid areas remain relatively warm. These concepts are important to our discussions of fog, lake effect, and air masses.

Diurnal temperature range is the earth's daily range in temperature. The earth receives its greatest amount of solar radiation at noon (sun time) and the least heating at sunset and sunrise. But, if the greatest heating is at noon, why doesn't maximum temperature occur at noon? Surface air temperature is also affected by the amount of reflected radiation. At noon, the earth is receiving its greatest heat, but it still is not warm enough to radiate as much energy as it is receiving. Therefore, surface air temperature continues to get warmer. Shortly after noon the earth gets still warmer, but begins to receive less solar radiation. Finally, incoming solar radiation exactly balances outgoing terrestrial radiation, and that is the time of maximum temperature. Maximum temperature, typically, occurs just before midafternoon.

During the night the surface receives no solar radiation, but it continues to radiate and become cooler. Just after sunrise, solar radiation begins again, but at a very low angle. Very little heating occurs while the surface continues to radiate and get cooler. Not long after sunrise, solar radiation again balances terrestrial radiation, and the surface begins to warm again. This is the time of minimum temperature.

Maximum and minimum temperatures affect aircraft performance and density altitude. Minimum temperature also plays a part in fog formation and the possibility of frost hazards.

The international standard for temperature measurement is Celsius. Winds aloft forecasts have always used this scale. Since July 1, 1996, with the adoption of the METAR (Aviation Routine Weather Report) weather code, surface temperatures for aviation in the United States, as well as the rest of the world, are reported in Celsius. Therefore, we will use this system. (Like the airspace reclassification in 1993 and the METAR/TAF codes in 1996, we might as well get used to using Celsius.) To this end the following benchmark temperatures should help us cope with the change from Fahrenheit (F) to Celsius (C).

−40°C =	−40°F	Eskimos stay in the igloo
−20°C =	−4°F	Chilly, even in North Dakota
−10°C =	14°F	Break out the gloves and ear muffs
0°C =	32°F	Water freezes
10°C =	50°F	Need a light jacket
20°C =	68°F	Room temperature
30°C =	86°F	Turn on the air conditioner
40°C =	104°F	We're having a heat wave

For those who still need a crutch, Appendix A, Conversion Tables, provides Celsius to Fahrenheit conversion.

A concept that is continually used is standard temperature. Standard temperature at sea level is 15°C. This value is nothing more than an average temperature taken from measurements all over the earth at sea level over a period of many years. Actual temperature for any location is rarely standard. The purpose of this value is to provide a common reference point.

Aircraft performance charts are based on the standard atmosphere, or more precisely, the international standard atmosphere (ISA). The standard *lapse rate*—decrease of temperature with height—is approximately 2°C per thousand feet. Temperature decreases to a value of −57°C at approximately 36,000 feet. Above 36,000 feet, in the standard atmosphere, temperature remains constant to about 66,000 feet, as illustrated in Fig. 1-3.

As mentioned, standard conditions rarely occur in the real world, and performance charts are based on standard conditions. Accommodations must be made for a nonstandard environment, most significantly a temperature correction. Manufacturers sometimes provide an ISA conversion with cruise power setting charts for high, low, and standard temperatures or simply note that performance is based on standard conditions. The aircraft doesn't understand any of this and performs based on the prevailing flight environment.

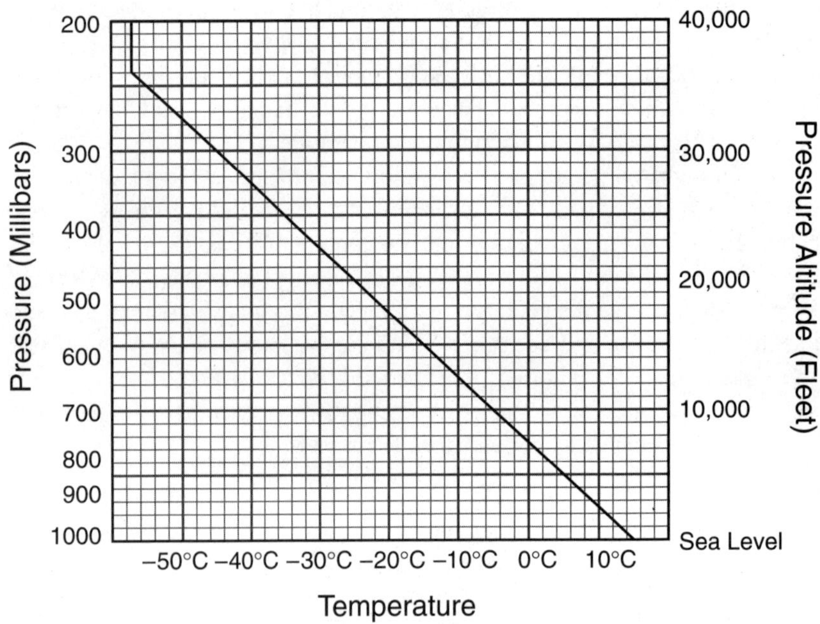

Fig. 1-3. *Aircraft performance charts are based on standard atmosphere, or more precisely, the international standard atmosphere (ISA).*

Pressure

Pressure is defined as force per unit area. For example, tires are inflated to a specified pressure, say 30 pounds per square inch. The atmosphere exerts pressure. At sea level atmospheric pressure is approximately 14.7 pounds per square inch. (We don't directly experience this pressure because our bodies exert an opposite and equal force. However, we do experience pressure changes with our ears, which must be cleared periodically with changes in pressure or altitude.)

Atmospheric pressure is the sum of all the air molecules above a specific point on the earth. Imagine a column of bricks. There are 10 bricks in the column, each weighing 1 pound. If we weigh the bottom of the stack, the scale reads 10 pounds. However, if we weigh the stack at the fifth brick, the scale reads 5 pounds; at the first brick, 1 pound. The atmosphere behaves in the same way.

For aviation purposes, we commonly relate atmospheric pressure to inches of mercury for example, the *altimeter setting, or pressure in millibars.* Inches of mercury (in. Hg), however, is not a direct expression of force per unit area. The international unit of atmospheric pressure is the hectopascal (hPa), which is equivalent to the millibar (mb).

So why do we measure pressure in inches of mercury? In 1643 an Italian, Evangelista Torricelli (1608–1647), invented an instrument for measuring atmospheric pressure—the mercurial barometer. Mercury was used because it was the heaviest liquid available that remains liquid at normal temperatures. (If water were used, the column would be 30 feet high!) Sea-level pressure balances a column of mercury approximately 30 inches high.

Like temperature, pressure has a standard. At sea level it is 29.92 in. Hg, or 1013.2 hPa (1013.2 mb). More about this when we discuss altimetry. And, guess what? For every level there is a standard temperature and pressure in the standard atmosphere.

In aviation weather we often refer to a constant pressure surface. We have already mentioned that the troposphere is higher at the equator than at the poles (see Fig. 1-1). Why? Air is warmer at the equator; therefore, molecules are farther apart. Columns of warm air are taller than columns of cold air. Let's take the 500-mb constant pressure surface. Since pressure is the sum of air molecules above the level where the measurement is taken, the 500-mb surface will, typically, be higher at southern latitudes in the northern hemisphere. Layers in the atmosphere are described in this manner. In aviation we most often refer to the 850-, 700-, 500-, 300-, and 200-millibar constant pressure surfaces. From Fig. 1-3 we see that these constant pressure surfaces occur at approximately 5,000, 10,000, 18,000, 30,000, and 39,000 feet, respectively.

The slope of the pressure surface, caused by horizontal temperature differences, determines approximate windspeed. Typically, the 500-mb slope is greater than the slope at 700 mb, and the 300-mb slope is greater than the slope at 500 mb. When pressure surface slope increases with height, windspeed increases with height. This is the general case in the troposphere. Winds are light or calm in areas of little or no horizontal temperature difference. And, in some cases, winds can decrease with height in the troposphere. This tends to occur within large high-pressure areas.

This information can be related to our discussion of the jet stream. Typically, the highest winds in the atmosphere occur with the jet stream, the area of greatest temperature difference.

The atmosphere, a gas, is compressible. That is, the weight of air molecules above compresses, or pushes together, the molecules below, as illustrated in Fig. 1-3. Notice that the pressure at sea level is approximately 1,000 mb. Pressure decreases with height, but not at a constant rate. Half the atmosphere by weight exists below 18,000 feet—the 500-mb level.

We have already discussed how pressure decreases with altitude. Atmospheric pressure also changes with the movement of weather systems. We have all heard the local weather forecaster talking about the movement of high- and low-pressure areas. So we have an atmosphere where pressure decreases with height and where pressure at the surface and aloft vary with latitude and weather systems.

Moisture

If we think of heat as the energy that drives the weather machine, then moisture, in the form of water vapor, is the fuel. Without moisture there would be no clouds or precipitation; in other words, there would be no weather as we know it.

Moisture in the atmosphere occurs in the form of ice crystals (a solid), water (a liquid), or water vapor (a gas). Clouds and precipitation are made up of ice crystals or liquid water.

Precipitation is any form of water (solid or liquid) that falls to the earth's surface. Water vapor is invisible, suspended in the air. The effects of water vapor are an important phenomena in meteorology and aviation weather.

The amount of water vapor in the atmosphere is measured in several ways. The term that most people are familiar with is *relative humidity*. A term often used in aviation is *dew point*.

Relative humidity is the ratio, expressed as a percentage, of the water vapor present in the air compared to the maximum amount of water vapor the air could hold at its present temperature. So, what does that mean?

Because air is a mixture of gases, it has no inherent capacity to hold moisture. The space available in a particular sample of air determines the amount of water vapor molecules it can hold. Since air molecules in warm air are farther apart, warm air can hold more water vapor than cold air. We use this principle many times in our discussion of weather-producing systems.

Figure 1-4 shows three air samples. The air samples have temperatures of 25°C, 15°C, and 5°C, respectively. Notice that each sample contains the same amount of water vapor. However, the warmer samples have the capacity to hold additional water in the form of vapor. The relative humidity of the warmer samples are thus lower than the cooler samples.

When relative humidity reaches 100%, the air is *saturated*, which means that the air can no longer hold any additional water in the form of vapor. This situation is illustrated by the 5°C sample in Fig. 1-4. Should any more water vapor be added or the air cooled

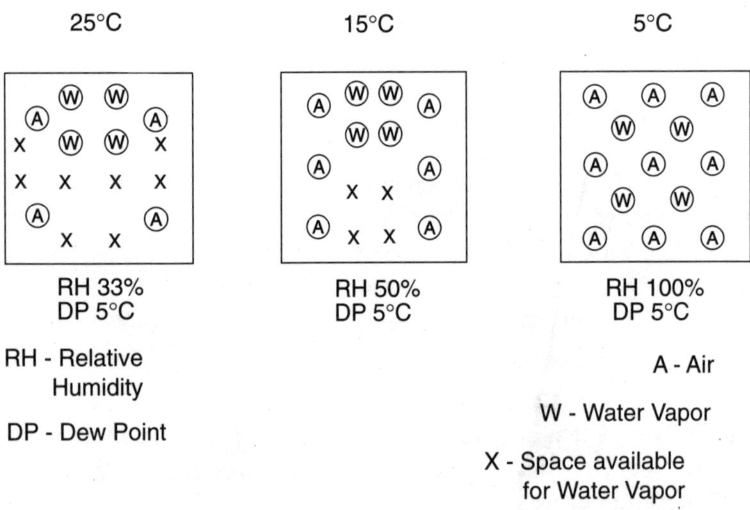

Fig. 1-4. *The warmer the air, the greater its capacity to hold water in the form of vapor.*

to a lower temperature, condensation will occur in the form of clouds or precipitation—visible moisture.

Dew point is the temperature to which air must be cooled, water vapor remaining constant, for air to become saturated. The dew point is usually available in METAR and SPECI (Aviation Selected Special Weather Report) reports, along with temperature, expressed in degrees Celsius. In Fig. 1-4, all samples have a dew point of 5°C. Therefore, the first sample would have to be cooled by 20°C and the second by 10°C to become saturated.

Energy, in the form of heat, is required to change water from a solid (ice) to liquid and from liquid to a gas (water vapor). The amount of heat exchanged (absorbed or released) is called *latent heat.*

Refer to Fig. 1-5. When ice melts, heat is absorbed. The heat required must be supplied from somewhere. It comes from the surrounding environment. In the process of freezing, heat is released to the surrounding environment.

When water evaporates (changes to water vapor) an even greater amount of heat is absorbed—without any change in temperature. The heat is supplied mostly by the liquid, with a smaller portion coming from the surrounding environment. We can all relate to the latent heat of evaporation. What happens when we jump out of a swimming pool? As the water evaporates, our skin cools. When water vapor condenses (changes to liquid), heat is transferred, or released, to the environment. A primary source of energy in the weather machine comes from the heat released from condensation—the latent heat of condensation.

Can water vapor change to ice without going through the liquid stage? Yes. And ice can evaporate, bypassing the liquid stage. This process is called *sublimation.* For example, an aircraft with structural ice loses the ice once out of the icing environment, even though the temperature remains below freezing. In the process of frost formation, water vapor in the atmosphere sublimates directly into ice. In this process of sublimation, a large amount of heat is absorbed or released.

Melting, freezing, evaporation, condensation, and sublimation are important factors in the weather machine. We constantly refer to this process in subsequent chapters.

In aviation we often use the term *freezing level.* It is actually a misnomer. Water melts at 0°C, but it can exist in the liquid state at temperatures well below freezing! Liquid water or water vapor that exists at temperatures below freezing is called *supercooled.* Because water can exist at temperatures below freezing, as supercooled cloud and water droplets, structural icing can occur.

Another important concept is the heat capacity of water vapor. We have already mentioned that water vapor is transparent to solar radiation but largely opaque to terrestrial radiation. That is, water vapor in the air absorbs terrestrial radiation and converts it to heat, raising the temperature of the air. Water vapor absorbs outgoing terrestrial radiation much as heat is trapped in a greenhouse—the greenhouse effect.

Density

Although temperature and pressure are important atmospheric properties, to understand the weather machine a thorough knowledge of density is second only to moisture. As we shall see, density is a key player in most weather phenomena.

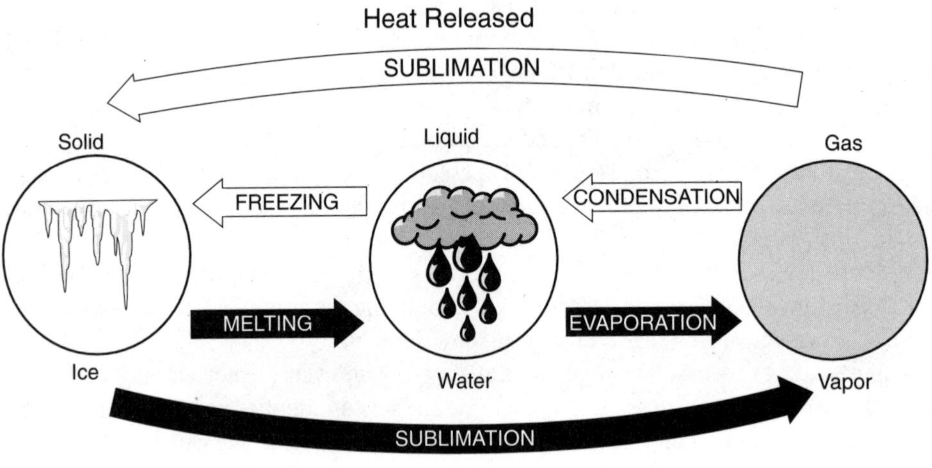

Fig. 1-5. *Energy, in the form of heat, is required to change water from a solid to liquid and from liquid to gas; the amount of heat exchanged is called latent heat.*

Density is the weight of air per unit volume, often expressed as "pounds per cubic foot" or "grams per cubic meter." Atmospheric density varies both horizontally and vertically, an important fact. At this point we can begin to put together some of the concepts we've discussed thus far and draw some practical relationships.

As we fly higher in the atmosphere, density decreases. As noted, density is weight, and we know the higher we go, the fewer molecules above; therefore, the air is less dense.

Atmospheric pressure also affects density. When temperature and humidity—moisture—remain constant, if pressure is higher than standard, density is higher; conversely, if pressure is lower than standard, density is lower.

Although pressure is an important factor in the density equation, temperature is often more significant. Typically, the higher the temperature, the lower the density of the air. Remembering that density is the weight of molecules, at higher temperatures there are fewer molecules per unit volume, pressure remaining constant. This concept can be seen in Fig. 1-4. There are fewer air molecules per unit volume in the warmer samples.

The final factor in air density is moisture. The higher the moisture content of the air, the lower its density. Why? Water molecules weigh less than air molecules. This is not a major factor, however. What we need to remember is that humid air is less dense than dry air.

The bottom line:

- Density has a direct relationship to pressure—as pressure decreases, density decreases.
- Density has an inverse relationship to temperature—as temperature increases, density decreases.

- Density has an inverse relationship to moisture (water vapor)—as moisture increases, density decreases.

ALTIMETRY, AIRSPEED, AND AIRCRAFT PERFORMANCE

The two primary flight instruments directly affected by air density are the altimeter and airspeed indicator. Air density also affects aircraft performance, including powerplant output. To safely and effectively use the aircraft, a pilot must understand how the atmosphere affects these instruments and the aircraft.

The following terms represent various altitude and airspeed references. Pilots must understand and be able to apply these terms to aircraft operations.

- Indicated altitude
- True altitude
- Pressure altitude
- Density altitude
- Indicated airspeed
- Calibrated airspeed
- True airspeed

Each of the preceding altitudes and airspeeds are defined and applied in the following sections.

Three flight instruments operate from air density. In addition to the altimeter and airspeed indicator, the vertical speed indicator operates off differences in air density. These instruments are connected to the aircraft's pitot-static system. Refer to Fig. 1-6. Ram air

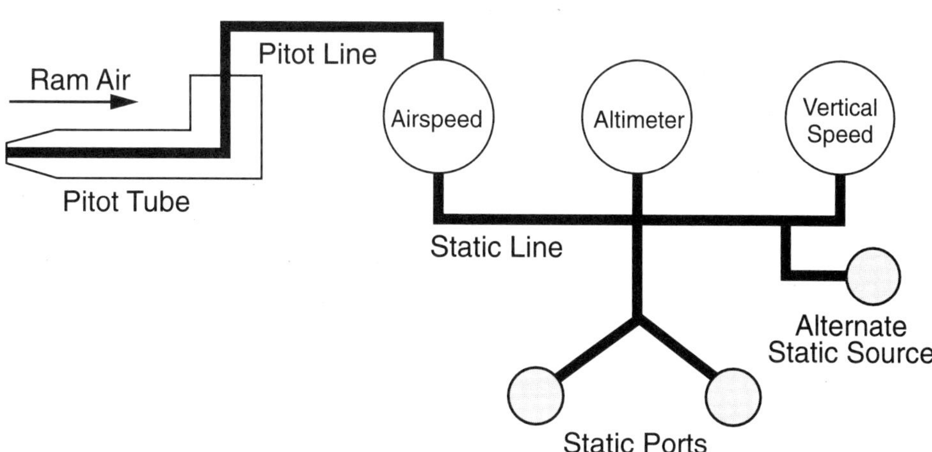

Fig. 1-6. *The pitot-static system uses changes in atmospheric pressure to generate instrument readings.*

pressure is connected to the airspeed indicator. The airspeed indicator measures the differential pressure between the ram air and static air pressure from the static ports. The altimeter and vertical speed indicators are also vented to ambient, or static, air pressure through the static ports, usually located on the side of the aircraft.

Many aircraft are equipped with an alternate static source should the static ports be blocked—iced over, for example. In an emergency, the alternate static source vents the static line to the cabin. If the alternate source is vented inside the aircraft, where static pressure is usually lower than outside static pressure, selection of the alternate source may result in the following instrument indications:

- Airspeed reads greater than normal.
- Altimeter reads higher than normal.
- Vertical speed indicates a momentary climb.

Airspeed and altitude read higher than normal because static pressure is lower than the environmental pressure. Vertical speed momentary indicates a climb because the instrument senses a momentary decrease in pressure, as if the aircraft were climbing into less dense air.

The Altimeter

The altimeter measures the height of the aircraft. Similar to an aneroid barometer, the altimeter measures changes in pressure as the aircraft climbs or descends. Figure 1-7 shows the aneroid wafers. As pressure decreases, the wafers expand. Through mechanical linkage, the change in altitude is reflected on the altimeter face. Conversely, when pressure increases, the wafers are compressed, and the altimeter indicates a descent.

The altimeter indicates correct altitude only under standard atmospheric conditions of pressure and temperature. For example, a pilot departs Oklahoma City under standard atmospheric conditions (29.92 in. Hg), climbs to 8500 feet, proceeds en route to Denver, and maintains 8500 feet on the altimeter. Let's say that at Denver atmospheric pressure is one inch lower than standard (28.92 in. Hg). As the aircraft proceeds west, pressure decreases. The aneroid wafers sense this decrease in pressure and expand, indicating a climb. The pilot, maintaining 8500 feet, is in a continuous slight descent. Arriving over Denver, the aircraft's actual altitude is only 7500 feet!

Since atmospheric conditions vary continuously, the altimeter must be adjustable for a nonstandard environment. Most altimeters are equipped with an altimeter setting window, sometimes known as the "Kollsman window" (Fig. 1-7). Paul Kollsman, a German-born aeronautical engineer, invented the method to correct the altimeter for nonstandard pressure in 1928, a major step in allowing pilots to fly solely by instruments. The altimeter setting window allows the pilot to adjust the instrument for nonstandard pressure using the altimeter set knob (Fig. 1-7). Typically altimeters can be adjusted for pressures from 28.10 to 31.00 in. Hg.

In the United States, Federal Aviation Regulations (FARs) require the pilot, when operating below 18,000 feet mean sea level (MSL), to fly with an altimeter setting of a

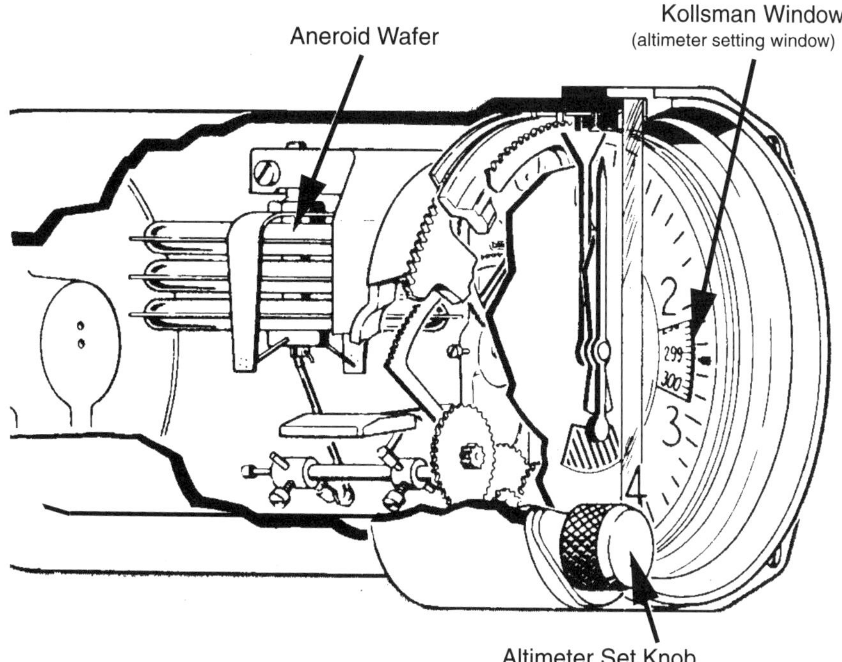

Fig. 1-7. *Similar to an aneroid barometer, the altimeter measures changes in pressure as the aircraft climbs or descends.*

station within 100 nautical miles (nm) of the aircraft. With the local altimeter setting in the window, the altimeter indicates altitude above MSL, called *indicated altitude.*

Nonstandard temperature affects the altimeter, although not to the same degree as pressure. Since warm air is less dense than cold air, the aneroid displays a lower altitude in cold air than in warm. For example, with a temperature 10° colder than standard and a pressure ½ inch below standard (worst-case scenario), the aircraft would be approximately 500 feet below indicated altitude. This condition has led to a sage saying when flying with an uncorrected altimeter: "When flying from high to low (or hot to cold), look out below!" Indicated altitude corrected for temperature is often called *true altitude.* However, the strict definition of true altitude is the actual or exact altitude above mean sea level.

For most practical purposes we can ignore temperature-induced altimeter error. Exceptions would be flying under visual flight rules (VFR) at low altitude—within 1000 feet of the surface—in extremely cold weather, especially at night or in low visibility conditions where visual contact with terrain cannot be maintained.

Pressure altitude is the altitude indicated on the altimeter when it is set to 29.92 in. Hg or 1013.2 mb. (When the altimeter setting equals 29.92 in. Hg, indicated altitude and pressure altitude are the same—standard conditions.) Altitudes on the ISA chart (see Fig. 1-3) are pressure altitudes. Pressure altitude is used when flying at or above 18,000 feet in the United States and in aircraft performance calculations to correct for nonstandard pressure.

Pilots using the local altimeter setting fly indicated altitude for levels through 17,500 feet and pressure altitude at 18,000 feet and above. Pressure altitude is used to eliminate station barometer errors and some altimeter instrument errors and to relieve air traffic controllers and pilots from having to reset the altimeter at frequent intervals due to the high speeds typically flown at and above 18,000 feet.

When pressure altitude is flown, altitudes are referred to as *Flight Level* (FL), for example, Flight Level one eight zero" (FL180 or a pressure altitude of 18,000 feet).

Under conditions of unusually low pressure, the lower flight levels (FL180 to FL200) may become unusable because pressure altitudes conflict with the higher available indicated altitudes (17,000 feet). Extremely high pressures, above 31.00 in. Hg, pose another problem. The altimeter cannot be set to the appropriate altimeter setting. Special procedures are then implemented. These procedures, along with lowest usable flight level, are contained in FARs and the *Aeronautical Information Manual* (AIM).

Pilots flying flight levels are actually flying a constant pressure surface. (Previously, we mentioned how the height of a constant pressure surface changes due to differences in temperature. Remember our axiom "when flying from hot to cold." Constant pressure surfaces are lower in cold air than in warm air.) The aircraft will climb or descend along with the height of the pressure surface. In the contiguous United States, terrain clearance is not a problem, even during periods of extremely low pressure and temperature. However, in other parts of the world, the transition between indicated altitude and pressure altitude may be different. The following codes are used internationally to define the appropriate altimeter setting:

- QNH
- QNE
- QFE

QNH represents indicated altitude—the pilot is flying an altitude above mean sea level. *QNE* is pressure altitude. We may occasionally see QFE. *QFE* is the altimeter setting used so that the altimeter reads zero when the aircraft is on the ground.

While flying in England (low, flat terrain), we occasionally used QFE. I explained to an English pilot that in the United States we used indicated altitude or QNH. He thought that was kind of silly, until I explained that it would be rather hard, in fact impossible, to crank 5000 feet out of the altimeter setting window for a landing at Denver, Colorado (elevation 5431 feet).

Winds-aloft forecasts use true altitude through 12,000 feet, so there will be a slight difference between the altitudes in the forecast (true altitude) and those flown by the pilot (indicated altitude). However, unless atmospheric pressure is extremely high or low, the difference is negligible. A variance of 1 inch of mercury from standard results in a 1000-foot altitude difference.

A final thought on altimeter settings. We were flying a Bonanza from Bakersfield, California, to Las Vegas, Nevada. A weather advisory was in effect for severe turbulence and there were strong, gusty surface winds in the Las Vegas area. For some reason, the Automatic Terminal Information Service (ATIS) at Las Vegas was unavailable. We contacted approach for clearance into Class B airspace and reported our altitude as 8500 feet. Approach reported our Mode C (altitude-reporting transponder) readout as 8000 feet! The aircraft's Mode C transponder reports pressure altitude. The air traffic control (ATC) computer adjusts this, based on the altimeter setting, to display indicated altitude to the controller. We had not updated our altimeter setting since leaving Bakersfield. There was a ½-inch difference at Las Vegas. The steep pressure gradient and lower pressure at Las Vegas was causing the strong winds and turbulence. From high to low, look out below!

In a previous section, we defined density. Air density is affected by atmospheric pressure, altitude, temperature, and to a minor degree humidity. *Density altitude* is pressure altitude corrected for nonstandard temperature. Density altitude affects aircraft performance, which is discussed later in this chapter.

Airspeed

Like the altimeter, air density affects the airspeed indicator. The airspeed indicator is a sensitive, differential pressure gauge. The instrument indicates the speed of the airplane through the air—not necessarily over the ground. The airspeed indicator measures the difference between the impact pressure of the air in the pitot tube and the static pressure of the ambient air surrounding the airplane, as shown in Fig. 1-6. A pointer on the face of the instrument registers the difference in pressure.

Indicated airspeed (IAS) is the direct instrument reading obtained from the indicator, uncorrected for installation and instrument error or variations in atmospheric density. *Calibrated airspeed* (CAS) is indicated airspeed corrected for installation and instrument error. These errors are usually greatest at low airspeeds, with flaps deployed. Normally, the manufacturer provides a table to convert indicated to calibrated airspeed for different speed ranges and flap settings.

True airspeed (TAS) is calibrated airspeed corrected for atmospheric density—pressure and temperature. Because of the normal decrease in pressure and temperature with altitude, for a given indicated airspeed, true airspeed increases as altitude increases. The pilot normally uses a flight computer to calculate true airspeed based on pressure altitude, temperature, and calibrated airspeed.

Aircraft Performance

All aircraft have limitations, including turbojet aircraft, which often fly at the edge of their performance envelopes. Nonstandard conditions can critically affect performance. The pilot's task is to determine aircraft performance based on actual or forecast conditions. This may require the information in Fig. 1-3, international standard atmosphere.

Certain performance charts require the pilot to determine temperature at altitude relative to ISA. For example, the forecast temperature over Reno at FL300 (30,000 feet) is −42°C. Figure 1-3 indicates that the standard temperature for that pressure altitude is

−45 °C. Certain flight computers can be used to determine ISA temperatures; these computers have a true altitude computation window. With the scales aligned (10 on the outer scale with 10 on the inner scale), standard temperature is read under pressure altitude. Under 30,000 feet pressure altitude, −45 appears. Therefore, the forecast temperature is ISA +3, or 3 degrees warmer than standard.

STRATEGIES

Now let's put the theory into practical application. Aircraft operations can be divided into takeoff, climb, cruise, descent, and landing.

High temperatures at high-altitude airports produce high density altitude. Atmospheric pressure and humidity are also factors; however, temperature and elevation are paramount. Surface temperature forecasts are not normally available, but maximum temperatures typically occur beginning by midmorning and continuing through late afternoon. Arrival and departure times must be planned based on aircraft performance.

Under certain conditions, the pilot can get the airplane off the ground only to be trapped in ground effect. *Ground effect* is the temporary gain in lift during flight at altitudes of about the wingspan of the airplane, due to compression of the air between the wing and the ground. If this is allowed to continue beyond the end of the runway, only one thing results—an airplane accident!

The following is a typical scenario: A pilot attempts to take off from a relatively short runway, during midday, with an aircraft at or above maximum gross takeoff weight. The aircraft accelerates and lifts off and may initially establish what the pilot perceives as a positive rate of climb. However, the aircraft then begins to settle, and with no runway remaining, it hits terrain. Pilots often blame downdrafts or windshear when weather conditions are conducive to neither.

I have flown Cessna 150s out of Bryce Canyon, Utah, (elevation 7586 feet), and Grand Canyon, Arizona (elevation 6606 feet), and a Cessna 210 out of Mammoth Lake, California (elevation 7128 feet)—density altitude 10,000 feet. There is no additional hazard in such operations as long as we calculate and do not attempt to exceed aircraft performance.

After calculating that the aircraft has sufficient performance for conditions, which is required by FARs, it's a good idea to determine an abort point on the runway. If the aircraft is not airborne and climbing out of ground effect, this point should allow the pilot to come to a safe stop on the remainder of the runway. Remember, aircraft performance data are based on brand-new airframe and engine and perfect pilot technique.

The pilot of a Beech A36 Bonanza attempted to depart Las Vegas, New Mexico from a 5000-foot strip, density altitude 8800 feet. After two aborted takeoffs, due to what the pilot reported as engine problems, a third try was attempted. The pilot tried to get the airplane airborne at the end of the runway but was unsuccessful. The airplane hit the ground and rolled off the runway. The airplane's pilot operating handbook (POH) indicated the takeoff distance should have been about 2000 feet. The pilot failed to determine the required takeoff distance and then failed to abort the takeoff at midfield when sufficient speed was not attained.

Normally, a pilot of a carbureted, fixed-pitch propeller airplane adjusts the mixture prior to takeoff for maximum engine RPM. Some pilots run the engine to maximum power during the runup and then adjust the mixture. (Cessna recommends this latter procedure when density altitude reaches 3000 feet. However, always check the POH for the airplane you're flying for recommended procedures.) I prefer to adjust the mixture to a rough setting at runup rpm and make final mixture adjustments during the initial takeoff roll. This procedure puts extra workload on the pilot during this critical phase but less stress on the airplane.

A pilot of a nonturbocharged, constant-speed propeller airplane uses the procedure described previously with the addition of leaning for maximum manifold pressure. The pilot must realize that manifold pressure reduces proportionally to density altitude. For example, with a density altitude of 6000 feet, the pilot could only expect 24 inches of manifold pressure at full throttle.

An advantage of a turbocharged engine is the development of sea level power at altitude. The turbo Cessna 182 POH recommends leaning to smooth engine operation. On airplanes equipped with a fuel flow indicator, for example, the Bonanza, the pilot can lean to density altitude takeoff fuel flow. This setting is a rough estimate, and smooth engine operation should always be maintained.

Multiengine pilots must consider that airport density altitude may exceed inoperative engine ceiling. For example, an early model nonturbocharged Cessna 310 has a single-engine ceiling of approximately 6800 feet, a Piper Aztec, 7500 feet, and a Piper Seminole, only 5000 feet. Remember, these altitudes apply to standard conditions. Density altitude in the summer, especially in the West, can often exceed these values.

Pilots cannot allow themselves to be lured into a false sense of security because the airport is at a relatively low elevation. The pilot of a Piper Saratoga, a turbocharged airplane, elected to depart a 1913-foot grass strip in Texas, field elevation less than 1000 feet, air temperature 34°C. The density altitude was approximately 3000 feet. The grass was wet and 4 to 6 inches tall. The aircraft struck power lines that were about 30 feet tall. The aircraft was slightly over gross weight. Based on the POH, with a dry runway, the aircraft required 1700 feet to clear a 50-foot obstacle. For a wet, sod runway, the distance required would be approximately 2300 feet!

Most density altitude accidents involve improper takeoff procedures. In nearly half the accidents the pilot attempted takeoff in an aircraft that exceeded the maximum gross weight. About one-third exceed the climb performance of the aircraft. Had these pilots calculated the aircraft's gross weight, takeoff distance, and climb performance, and not attempted to exceed the aircraft's performance, all of these accidents could have been prevented.

Climb and cruise performance are also affected by density altitude. Take, for example, the following cases.

We had filed an IFR flight plan from Lancaster's Fox Field in the Mojave Desert to Ontario, California. I requested 7000 feet and planned to go through the San Fernando Valley because of lower minimum altitudes. The clearance came back, "cleared via the Cajon two arrival; climb and maintain 11,000." The surface temperature was 30°C and the Cessna 150 was not going to 11,000 that day! After negotiating with a rather perturbed ground controller, I received my requested routing. Pilots must know their aircraft's performance and not allow ATC, or anyone else for that matter, to push them into an untenable—in this case unobtainable—position.

Let's take another example. The average temperature from Oakland to South Lake Tahoe at 13,500 feet was −04°C. Figure 1-3 indicates that the standard temperature for 13,500 feet is −10°C. The forecast temperature was 6° warmer, which is above standard. The air was less dense than standard, so aircraft performance would be less than performance charts advertise. Based on these conditions, density altitude was about 15,000 feet. My 1966 Cessna 150 had a book ceiling of 12,650 feet. We planned to traverse the 9941-foot Tioga Pass in California's Sierra Nevada range. The winds were out of the northeast at only 10 knots, resulting in a slight downdraft from the wind flowing up the east slopes and down the west slopes as we approached the pass. Temperature was slightly above standard, and in combination with the wind, the aircraft wouldn't climb out of 9500 feet. We had to proceed north along the west slopes of the mountains to Ebbett's Pass at 8732 feet, where we were able to safely cross the mountains.

You should gain the required altitude prior to reaching the pass or crest of the mountains, and with sufficient room to make a comfortable course reversal if required. One technique is to approach the ridge line at a 45° angle. If the ridge cannot be cleared, the pilot usually only has to make a 45° turn away from the mountains to reach lower terrain. How can we tell if we'll clear the crest? If we're above the crest, the terrain beyond will appear to be descending in relation to the crest of the mountains.

An aircraft with an advertised service ceiling of 13,100 feet is based on standard conditions. Differences are usually not significant unless the pilot is operating at the limit of the aircraft's performance. Unfortunately, this occurs every year with pilots who attempt to cross the Sierra Nevada or Rocky Mountains in conditions well above standard. Some pilots can't understand why an aircraft with a service ceiling of 13,100 feet can't climb above 12,000 feet with an outside air temperature of 0°C. Density altitude at 12,000 feet and temperature 0°C is 13,100 feet! And this is an ideal case. A runout engine, poor leaning technique, over-gross weight, and the possibility of turbulence and downdrafts further decrease performance. Some have mused that you can walk across the Rockies and Sierra Nevada on the wreckage of Cessna 172s and Piper Cherokees. You can't fool Mother Nature; attempts can be fatal.

Descents are normally not a problem, but, don't forget to richen the mixture, if required.

A pilot should use the same indicated airspeed values for high density altitude approach and landing as those used at sea level. Why? Instrument indications are based on air density. But airplane speed over the ground (ground speed) will be higher. This is one reason for longer ground rolls at higher density altitudes. Multiengine pilots especially should know density altitude prior to landing. If a go-around should be required, the pilot should know if the airplane is above its single-engine ceiling.

The bottom line: Calculate existing density altitude, and don't expect the aircraft to exceed its design limitations. Most flight computers have density altitude functions. Various tables and charts have been developed for this purpose. Appendix A, "Tables and Graphs," contains a density altitude chart. There is no excuse for not obtaining and applying this information.

UPDATING WEATHER EN ROUTE

Updating weather en route begins with a complete preflight briefing. In fact, many pilots follow weather trends for several days before a planned flight. In this way they get a feel for the weather. Without this background knowledge, we cannot fully apply updated weather en route. For example, en route we obtain information that our destination is below our minimums. The next question is: Why? It is due to the delayed improvement of stratus and fog, the faster-than-forecast advance of a frontal system, or the approach of a hurricane? We need to know to develop a sound alternate plan.

However, a pilot's responsibility does not end with an understanding of weather and forecasts and a complete preflight briefing. Due to the dynamic character of the atmosphere, data must be continually updated. Surprisingly, many pilots have not been taught the importance of updating weather reports and forecasts. Failure to exercise this pilot-in-command prerogative, as we'll see, can have disastrous results.

Up-to-date and accurate information and a knowledge of how to apply it are the keys to intelligent cockpit weather decisions. There are four sources of information:

- The pilot
- Flight service stations (FSSs)
- En route flight advisory service (Flight Watch)
- Air traffic control facilities—centers and towers

The pilot is a key source through direct observation and analysis of aircraft instruments. Often the pilot's observations are the only means of evaluating cockpit weather conditions, especially cloud types, and winds and temperatures aloft. To successfully apply the information requires a sound knowledge of aviation weather phenomena and theory, a major theme of this book.

FAA flight service stations are a primary source of weather information. All of the information available during the preflight weather briefing is accessible in the cockpit through FSS communications outlets. This information includes current observations, pilot weather reports (pireps), radar reports, satellite imagery, and, often, update forecasts.

Selected FSSs provide a continuous broadcast of weather advisories and urgent pireps over selected radio navigation aids (VOR). This service is known as hazardous inflight weather advisory service (HIWAS). When a weather advisory affects an area within 150 miles of a HIWAS outlet, an alert is broadcast once on all frequencies, except Flight Watch and emergency.

En route flight advisory service (EFAS), radio call "Flight Watch," is a special function provided by selected flight service stations. Flight Watch provides meteorological information for the phase of flight that begins after climbout and ends with descent to land. Flight Watch is a central point for the collection and dissemination of pilot weather reports. The effectiveness of Flight Watch to a large degree depends on this two-way exchange of information.

The purpose of Flight Watch is to provide pilots with timely and meaningful weather advisories. The preflight briefing provides current and forecast conditions at the time of the briefing. This information should be updated routinely en route; the airlines do it, often through Flight Watch. Are updated reports consistent with the forecast? If not, why? Flight Watch specialists are in an excellent position to detect forecast variance. Whether the forecast was incorrect or conditions are changing faster or slower than forecast, the pilot needs to known and plan accordingly. Flight Watch is in the best position to provide the latest information and suggest possible alternatives.

Updates must be obtained far enough in advance to be acted upon effectively. Action must be taken before critical weather is encountered or fuel runs low. Arriving over a destination that has not improved as forecast or has deteriorated is folly. At the first sign of unforecast conditions, Flight Watch should be consulted and, if necessary, an alternate plan developed. This might mean an additional landing en route, which is eminently preferable to a terrifying flight or an aircraft accident.

Flight Watch is not intended for flight plan services, position reports, initial or outlook briefings, aeronautical information, or single or random weather reports or forecasts. This information should be obtained through normal FSS communication channels.

Flight Watch communications below 18,000 feet are on the common frequency of 122.0. Since Flight Watch has a number of transmitter sites on the common frequency, it is very important for pilots to state the aircraft's position on initial contact. For example, "Oakland Flight Watch, Cessna seven three five two juliett, Lake Tahoe, over." If you're not sure of which Flight Watch you're calling, just call Flight Watch and state the aircraft's position, "Flight Watch, Champ four three three zero charlie, Big Sur, over." (OK, that was the commercial. I would be severely criticized by my colleges at Flight Watch if I didn't mention it. The pilot's failure to provide position on initial contact is the biggest complaint from Flight Watch specialists.)

ATC specialists at towers and centers are another source of cockpit weather information. Many towers provide automatic terminal information service (ATIS) with local weather information and weather advisories. Radar facilities can assist with weather avoidance, but that's not their primary job and their radar is not designed specifically to detect weather.

Finally, there is the human factor. When, as pilots, do we say no and call it a day? I teach, or maybe preach, that the first time the thought occurs, "Should I really be here?" or "Maybe I should turn around" is a red flag to take positive action. Don't push the weather, your aircraft, or yourself; turn around and wait it out. As a friend, and excellent pilot, puts it: "There is never any reason that we absolutely have to be there."

2
Ever-changing weather

NOW WE CAN BUILD ON THE KNOWLEDGE OF CHAP. 1. A MAJOR THEME of that chapter was the standard atmosphere. As we know in the real world, the atmosphere is rarely standard.

A dominant factor in the weather is stability or lack thereof. Stability is a major element in vertical motion. What happens when air is forced upward? It cools; but at what rate? It's rarely standard. We discuss and apply the different types of atmospheric stability: absolute and neutral stability and absolute and conditional instability. These processes occur in the real world and have significant effects on our flying activity.

Vertical motion is a primary weather producer. Clouds, weather, and aviation weather hazards are closely related to vertical motion. We introduce eight vertical motion mechanisms, any one of which, under the right conditions, is capable of initiating thunderstorms and other hazards. In subsequent chapters we refine and expand our knowledge of these phenomena. The bottom line: Both stability and vertical motion are important to our understanding of weather phenomena.

Since low ceilings and visibilities are directly related to stability and vertical motion, we include a discussion of these phenomena. Also included are the hazards of sand and volcanic ash. Heavy rain, snow, freezing rain, and hail present specific aviation hazards. Each is reviewed.

The chapter concludes with strategies to avoid the dangers of low ceilings and visibilities. As we'll see, the IFR pilot is not immune to these hazards.

LAPSE RATE

The *Glossary of Weather and Climate* (American Meteorology Society, 1996) defines lapse rate as the decrease of temperature with height. In Chapter 1 we discussed the standard lapse rate in the troposphere as approximately 2°C per thousand feet, as illustrated in Fig. 1-3. Well, what if the atmosphere isn't standard? (Here is where we apply our discussion of the change of state of water from Fig. 1-5.)

The lapse rate of dry air is 3°C per 1000 feet. That is the rate at which unsaturated air cools as it is forced upward—known as the *dry adiabatic lapse rate.* Conversely, when unsaturated air is forced downward, it heats at the rate of 3°C per 1000 feet. But what about saturated air? The moist or saturated lapse rate varies from between 1 and 2 degrees per 1000 feet. To understand how all this works, we need the process of adiabatic expansion and compression.

From Chapter 1 we know that atmospheric pressure is the weight of air above, which includes water vapor. What happens if air is forced upward? To understand we need the concept of an *air parcel.* A parcel of air is a small volume of air arbitrarily selected for study. It retains its composition; that is, it does not mix with the surrounding air. It responds to all meteorological processes. For example, it can expand or be compressed—thus temperature can change. Its moisture remains constant as long as its temperature allows moisture to remain in the vapor state (dew point remains lower than temperature). As an example, uneven surface heating causes parcels of air near the surface to be heated and rise. (This concept is illustrated in Fig. 3-8. Each cloud results from a parcel of air that has been lifted from the surface.) This process causes thermal turbulence. If the concept of an air parcel is not clear now, it will be by the end of this section.

As a parcel of air is forced upward, pressure decreases and the parcel expands—volume increases. Expansion pulls apart the molecules. Energy is required to keep the molecules apart. The energy comes from the gas itself in the form of a temperature decrease. If the parcel were to return to its starting point, the reverse process would occur. As pressure increases, temperature increases and volume decreases. This is the *adiabatic process,* defined as the internal changes within a body of gas during expansion or compression when no energy is added or removed.

In the general circulation, many comparatively large areas have gradually ascending or descending air. Moving air masses may be forced upward by sloping terrain. Large bodies of warm air often override colder air and are forced aloft by the cold air. In all these processes, the vertically moving air undergoes adiabatic changes.

Changes follow definite universal gas laws. The mathematical computation of the process is complex and cumbersome. However, we can illustrate the process graphically. Several charts have been developed for this purpose. For illustration, we will use the adiabatic chart beginning with Fig. 2-1.

Like the standard atmosphere chart in Fig. 1-3, the adiabatic chart shows temperature on the horizontal axis and pressure on the vertical, millibars on the left, pressure altitude

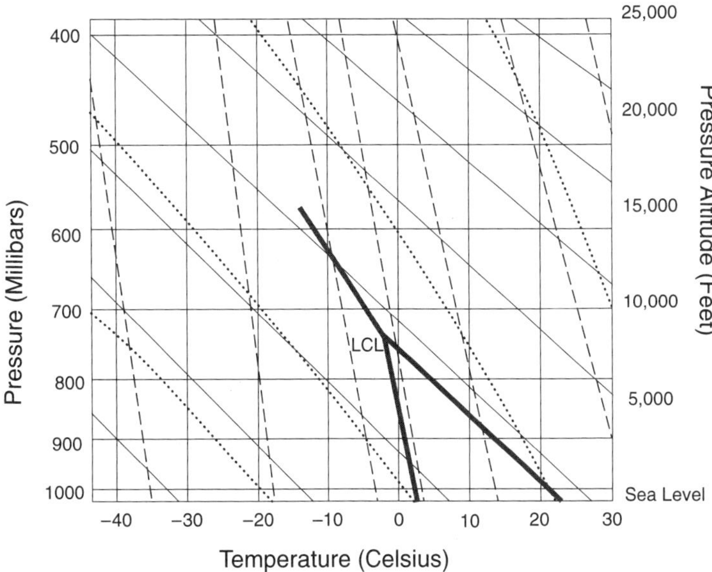

Fig. 2-1. *The adiabatic chart allows us to study the behavior of a lifted parcel of air.*

on the right. The solid diagonal lines represent dry adiabats. Dashed lines are mixing ratio—a measurement of water vapor content. Dotted lines show the saturated adiabats. We'll define all of these in a moment.

Refer to Fig. 2-1. Let's take a sea level parcel of air with a temperature of 22°C and a dewpoint of 3°C. If we lift the parcel, it will cool at the rate of 3 degrees per 1000 feet, along the dry adiabats (diagonal lines). As the parcel rises, the dewpoint temperature decreases along the mixing ratio (dashed) lines. When a parcel is lifted, the actual amount of water vapor does not change. However, its relative humidity increases and the dewpoint temperature changes, but at a much slower rate than the decrease in air temperature.

When unsaturated air rises, why does its temperature–dewpoint spread decrease and humidity increase? The temperature of a rising air parcel decreases due to expansion. The dewpoint of unsaturated, rising air remains approximately the same. This is true because the water vapor content of the air remains constant. Dewpoint temperature reflects the amount of water vapor in the air. Dewpoint temperature of unsaturated air actually decreases slightly as the air rises to a lower pressure, but the change is less than 1°C per 1000 feet. For our discussions, we will assume the dewpoint remains constant as long as the air remains unsaturated.

Since the temperature–dewpoint spread of rising air decreases, then, by definition, the air's relative humidity must increase. Conversely, the reverse process is true for descending air. The temperature of descending air increases while the dewpoint remains approximately constant. Thus the temperature-dewpoint spread of sinking air increases, which means relative humidity decreases.

Where the dry adiabat and mixing ratio lines intersect, the temperature and dewpoint are the same—the parcel has reached saturation. In the example, saturation occurs at 8000 feet. Our parcel has cooled to saturation due to expansion as it moved upward in the atmosphere, and yet no heat has been added or removed from the parcel. This intersection is known as the lifted condensation level (LCL). Relative humidity is 100%.

This process can be used to estimate the base of convective clouds. When a cloud layer is produced by surface heating, bases can be approximated by applying the following formula:

$$T - Td \times 4 \times 100 = \text{cloud base above ground level (agl)}$$

where T = temperature in degrees Celsius,
Td = dewpoint in degrees Celsius,
and 4 and 100 are constants.

For example, the temperature and dewpoint on the METAR are 15/05. Temperature minus dewpoint (15°C − 5°C) equals 10°C; apply the formula $10 \times 4 \times 100 = 4000$, and the approximate cloud bases are 4000 feet agl.

As another example, use this observation for Denver: METAR KDEN 182350Z ...SCT090 SCT150 BKN250 33/08... Applying the formula to the Denver metar, the cloud base works out to be approximately 10,000 feet agl ($33 - 08 \times 4 \times 100 = 10{,}000$). This is consistent with the observer's report of 9000 scattered. An article in a popular aviation magazine claimed this procedure could be used for any cloud layer. That's not true. It's important to remember the procedure only applies to convective clouds produced by surface heating—air lifted adiabatically.

To continue our discussion, we must employ the final set of lines on the chart, the saturated adiabats. Remember that condensation is a warming process—heat is released. This adds heat to the air and partially offsets adiabatic cooling due to expansion. As we continue to lift the parcel above saturation, it no longer cools at the dry adiabatic rate, because it absorbs the heat produced by condensation. Therefore, it cools at a slower rate. Cooling is now along the saturated adiabats (dotted lines).

Vertical motion also affects the moisture content of saturated air. Saturated air that rises must have its water vapor content decrease. This decrease occurs because the lowering temperature means its capacity to hold water vapor decreases. Where does the water vapor go? Water vapor must change from a gas within the air to water, in either a liquid or solid form, depending of the temperature of the air. Visible water condenses in the form of clouds and precipitation.

If we continue to lift the parcel, it cools at the saturated adiabatic lapse rate. In the example, if the parcel is lifted to 15,000 feet, its temperature will be −13°C. The temperature and dewpoint are the same; relative humidity remains 100%. However, moisture has decreased. Where has the moisture gone? Precipitation. Precipitation is the process by which liquid water is removed.

When saturated air sinks, but contains no visible moisture, the sinking motion is at the dry adiabatic rate. As the air descends, relative humidity decreases.

If visible water in the form of clouds or precipitation is present, the relative humidity of the descending air tends to decrease. The decrease is at the saturated adiabatic rate rather than the dry adiabatic rate. As relative humidity decreases, water particles evaporate. The evaporation increases the air's dewpoint. The latent heat loss by the air due to evaporation slows the temperature increase. The slower rate of increase of temperature–dewpoint spread means a slower decrease in relative humidity. In some instances with the presence of liquid droplets or precipitation, it is possible that the relative humidity remains nearly saturated. The amount of visible water in the form of clouds or precipitation decreases due to evaporation within the sinking air. Thus precipitation tends to diminish and clouds tend to dissipate.

What happens if we return our parcel to sea level? Refer to Fig. 2-2. We'll assume that our parcel is saturated but contains no visible moisture. The air warms at the dry adiabatic lapse rate of 3°C per 1000 feet. At sea level its temperature will be 30°C. If we follow the mixing ratio (dashed) lines, its dewpoint is −7°C. The air is hot and dry. This process explains some major meteorological phenomena, for example, the Santa Ana conditions of Southern California and the Chinook that develops along the eastern slopes of the Rockies. In fact, on rare occasions a phenomenon known as a "heat burst" is produced by a thunderstorm. The temperature rapidly but briefly jumps to extreme values. Near Kopperl, Texas, in 1966 the temperature was estimated at 60°C. It was thought to be caused by adiabatic warming in the extreme downdrafts associated with storm. More about this later.

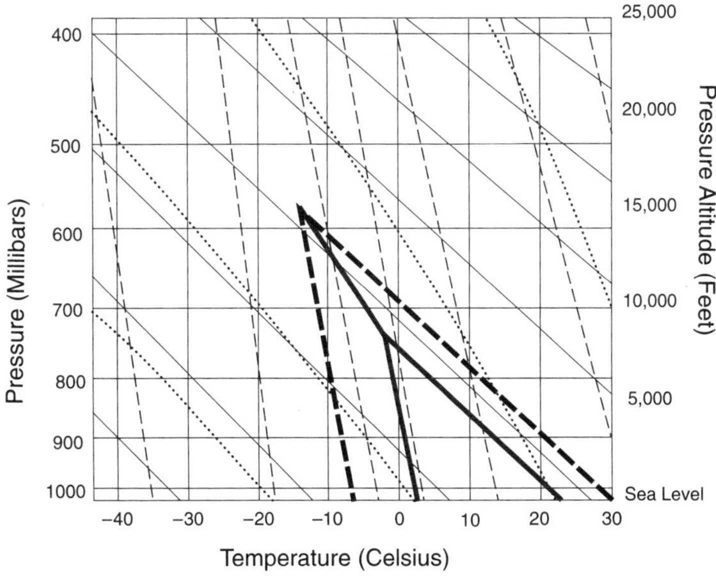

Fig. 2-2. *When a lifted parcel is returned to the surface, its properties change—it becomes warmer and dryer.*

Stability

Before we proceed we need to understand atmospheric stability. Stability is defined as the ability of a mass of air to remain in equilibrium—its ability to resist displacement from its initial position. If we move a parcel of air, and then remove the lifting mechanism, one of several things occurs: The parcel tends to return to its original position; the parcel continues to rise without any additional lifting force; the parcel initially resists upward displacement to a certain point, where it then spontaneously continues upward; or the parcel remains at the level where the external force ceases. These processes are known as

- Absolute stability
- Absolute instability
- Conditional instability
- Neutral stability

To explain how stability works, we return to the adiabatic chart where we can compare properties of a lifted parcel to the actual or existing lapse rate, which from now on we refer to as lapse rate.

A parcel is *absolutely stable* when it resists vertical displacement whether saturated or unsaturated. This condition is illustrated in Fig. 2-3. The lapse rate is shown as the large, solid black line. Notice that if a parcel is forced upward, it cools at the dry adiabatic rate when unsaturated and then the saturated adiabatic rate. In either case, it remains cooler than the surrounding air. As a matter of fact, as long as the existing lapse rate remains in the gray area of Fig. 2-3, the parcel is absolutely stable. Since the parcel is cooler (denser) than the surrounding air, it wants to sink. Thus, vertical motion is impossible, unless it is caused by an external force.

A stable lapse rate may exhibit the properties of an isothermal layer or inversion. Both are illustrated in Fig. 2-3. That portion of the sounding between the 700 mb and 600 mb levels is isothermal. That is, the lapse rate is constant. The lapse rate between the 600 mb and 500 mb levels is an inversion—temperature increases with altitude. An inversion strongly resists lifting and in many cases puts a "cap" on weather.

Air is *absolutely unstable* when vertical displacement of a parcel within the layer is spontaneous, whether saturated or unsaturated (Fig. 2-4). Again, the lapse rate is shown as the large, solid black line. If a parcel is lifted, its temperature is always warmer than the surrounding air. The cooler, denser air surrounding the parcel forces it upward. Vertical motion is spontaneous, and the layer is absolutely unstable. If in fact, as long as the existing lapse rate remains in the gray shaded area, a lifted parcel is absolutely unstable.

Let's look at the third case, *conditional instability.* Conditional instability refers to the structure of a column of air that produces free convection of a parcel as a result of it becoming saturated when forced upward. *Free convection* means that once saturation occurs, upward movement continues spontaneously. This is the *level of free convection* (LFC), as shown in Fig. 2-5. A lifted parcel is stable to the point where saturation occurs. Below the lifted condensation level the parcel exhibits absolute stability. However, upon saturation, upward displacement becomes spontaneous. This occurs when the lapse rate lies within the gray shaded area of Fig. 2-5. The parcel becomes unstable on the condition

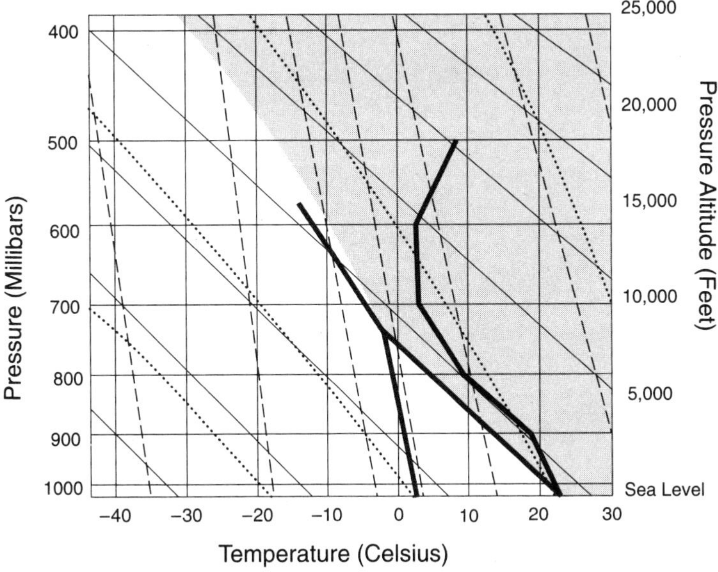

Fig. 2-3. *A parcel is absolutely stable when it resists vertical displacement, whether saturated or unsaturated.*

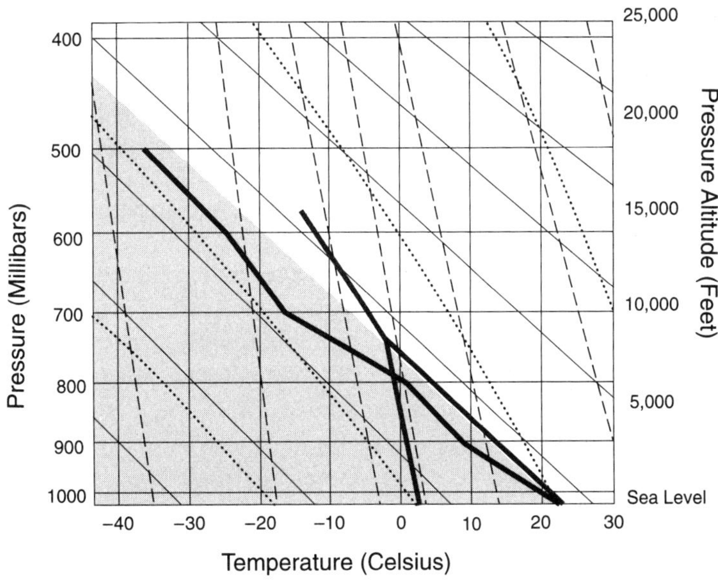

Fig. 2-4. *Air is absolutely unstable when vertical displacement of a parcel within the layer is spontaneous, whether saturated or unsaturated.*

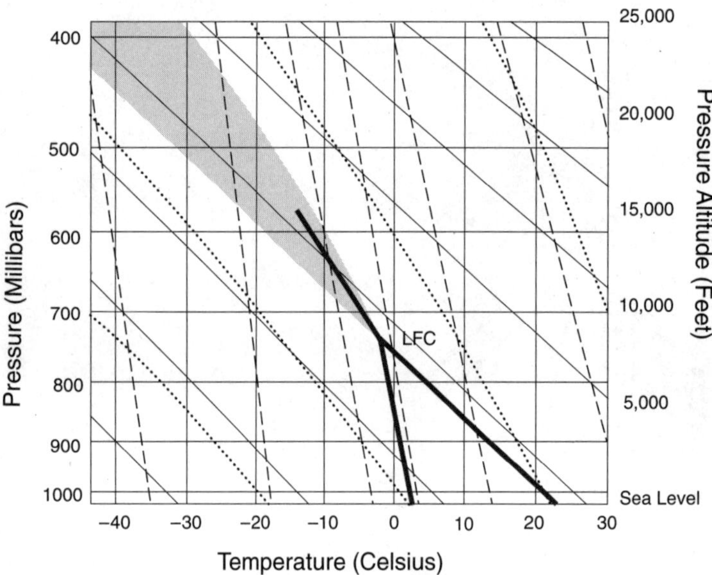

Fig. 2-5. *Conditional instability refers to the structure of a column of air that produces free convection of a parcel as a result of its becoming saturated when forced upward.*

that it reaches saturation. High moisture content in low levels and dry air aloft favor instability. Conversely, dry air in low levels and high moisture content aloft favor stability.

The concepts of absolute stability and absolute instability are relatively straight forward. Conditional stability is more complex and includes many subclassifications. It depends not only on temperature, but also on water vapor distribution.

When a parcel is displaced and remains at rest—even when the displacing force ceases—the layer is *neutrally stable.* For a layer of unsaturated air to be neutrally stable, its lapse rate must be equal to the dry adiabatic rate. For a layer of saturated air to be neutrally stable, its lapse rate must be equal the moist adiabatic rate.

Table 2-1 contains general characteristics of stable, conditionally unstable, and unstable air. When ceilings exist in stable air, they tend to be low, associated with stratiform clouds and fog. The stable air typically produces poor visibility. Flight conditions tend to be smooth. Any precipitation is typically light to moderate and steady. Conditional instability tends to exhibit the characteristics of stable air below the lifted condensation level and unstable air above. Precipitation is steady and light to moderate in intensity, with embedded areas of heavy showers. Ceilings and visibility are typically good in unstable air. Flight conditions are turbulence. Cloud types tend to be cumuliform, with showery, heavy precipitation. We build on these characteristics throughout the book.

VERTICAL MOTION IN THE ATMOSPHERE

In the previous section, we discussed stability and what happens when a parcel of air is lifted. A logical question is what atmospheric processes can lift a parcel? An understanding of vertical motion (both upward and downward) explains many weather phenomena and aviation weather hazards. In this section we briefly explore the methods employed by nature to move air and its properties vertically. We build on this section throughout the book.

Vertical motion can be produced, enhanced, or dampened by one or all of the following:

- Convergence and divergence
- Frontal lift
- Dry line
- Vorticity
- Upslope
- Pressure systems
- Convection
- Warm- and cold-air advection

Convergence refers to an inflow or squeezing of the air. In the horizontal, when air flowing into an area is greater than the outflow, the air literally piles up. Since the ground prevents the air from going downward, there is only one way left for it to go—up. The bottom line: An area of convergence is an area of rising air. Convergence can occur aloft over dense, cold air and is not necessarily confined to a layer bounded by the surface. When moisture is adequate and convergence is great enough, condensation occurs.

Convergence occurs along surface low pressure troughs and at the center of low pressure areas. An area of strong winds blowing into an area of lighter winds causes wind-speed convergence. When occurring near the surface, the result is a region of rising air.

Divergence is the opposite of convergence. Downward motion (subsidence) of air causes it to spread out at the earth's surface. From our knowledge of lapse rate, we can see that divergence is a drying and stabilizing process. Like convergence, divergence may occur in a layer aloft not extending to the ground. Divergence occurs along high

Table 2-1. Characteristics of Atmospheric Stability

Stable	**Conditionally unstable**	**Unstable**
Ceiling poor	Ceiling poor	Ceiling good
Visibility poor	Visibility poor	Visibility good
Smooth	Smooth below/turbulent above	Turbulent
Stratiform	Stratiform/embedded cumuliform	Cumuliform
Precipitation steady	Steady/embedded showers	Precipitation showery

pressure ridges and at the center of high pressure areas. When air near the surface blows from an area of light winds into an area of stronger winds, windspeed divergence results.

Areas of convergence and divergence can be located on surface analysis and 850- and 700-mb constant pressure charts.

Air masses of different properties (temperature and moisture) do not tend to mix. Differences in temperature, humidity, and wind may change rapidly over short distances. Where there are temperature and moisture differences, there is a difference in density. This zone of rapid change separating the two air masses is a *frontal zone,* more commonly referred to as a *front.* In these zones, the less-dense air is lifted, causing vertical motion in the atmosphere. The type of weather produced is dependent on the stability of the atmosphere.

A *dry line,* or temperature–dewpoint front, marks the boundary between moist, warm air from the Gulf of Mexico and dry, hot air from the southwestern United States. Dry lines usually develop in New Mexico, Texas, and Oklahoma during the summer months. Since the moist air from the Gulf is less dense than the dry, hot desert air, it is forced aloft. If the air mass is unstable, thunderstorms and tornadoes develop along the boundary. Areas of frontal lift and dry lines can be located on the surface analysis chart.

Anything that spins has vorticity, which includes the earth. *Vorticity* is a mathematical term that refers to the tendency of the air to spin; the faster that air spins, the greater its vorticity. A parcel of air that spins counterclockwise (cyclonically) has positive vorticity; a parcel of air that spins clockwise (anticyclonically) has negative vorticity.

Air moving through a ridge, spinning clockwise, gains anticyclonic relative vorticity. Air moving through a trough, spinning counterclockwise, gains cyclonic relative vorticity; therefore, there tends to be downward vertical motion in ridge-to-trough flow, and upward vertical motion in trough-to-ridge flow. More about this later.

Orographic is a term used to describe the effects caused by terrain, especially mountains. An orographic effect is upslope and downslope. The air can also take on the characteristics of the terrain through the process of conduction. That is, the air can absorb heat or moisture by direct contact with the surface. Air moving up a slope rises and tends to cool; air moving down a slope sinks and tends to warm. Areas of upslope can be determined from the surface analysis and weather depiction charts, with a knowledge of terrain.

Circulation around low pressure produces upward vertical motion; around high pressure, downward vertical motion. Typically, but not always, low pressure means poor weather and high pressure good weather. Low and high pressure areas are depicted on the surface analysis chart.

Upward vertical motion occurs in low pressure troughs (usually referred to as a trough) and surface lows due to convergence. Conversely, downward vertical motion occurs in high pressure ridges (usually referred to as a ridge) and surface highs caused by divergence.

Vertical motion also occurs at higher levels. For example, air moving from an upper level trough to ridge produces upward vertical motion; air moving from an upper ridge to trough produces downward vertical motion. Upper-level pressure patterns are depicted on constant pressure charts.

Atmospheric convection is the transport of a property vertically. This was mentioned in our discussion of stability, specifically the level of free convection. For our purposes, near the surface convection is caused by surface heating. Surface heating, and the resulting convection, is a primary vertical motion producer. In later chapters we see how this process affects the weather.

Advection is a term used to describe the movement of an atmospheric property from one region to another. Temperature is one property that can be advected. From the surface to about 10,000 feet, warm-air advection produces upward vertical motion and cold-air advection produces downward vertical motion. As warmer, less-dense air moves into an area, it tends to rise. Warm-air advection causes surface pressures to fall, resulting in convergence and upward vertical motion. When cooler, denser air moves into an area, it tends to sink. Cold-air advection causes surface pressures to rise, resulting in divergence and downward vertical motion. Therefore, warm-air advection destabilizes conditions, whereas cold-air advection tends to stabilize the weather at and near the surface.

Above the 500-mb level the opposite occurs. Cold-air advection destabilizes conditions, and warm-air advection stabilizes the atmosphere. Cold-air advection above the 500-mb level decreases the lapse rate, enhancing any convective activity that might develop. Conversely, warm-air advection aloft stabilizes the atmosphere by increasing the lapse rate, thus retarding any convection.

Areas of low-level cold- and warm-air advection can be determined from the 850- and 700-mb constant pressure charts. Areas of high-level cold- and warm-air advection can be determined from the 300- and 200-mb constant pressure charts.

LOW CEILINGS AND VISIBILITIES

We now have enough background to apply the "theory" to actual flight situations. We begin with low ceilings and visibility, which are significant hazards, especially to the pilots restricted to visual flight rules. A VFR-only pilot needs a natural horizon or ground contact to maintain control of the aircraft. Even though the private pilot curriculum includes basic, or emergency, instrument flying, the training is not sufficient for any length of time in instrument meteorological conditions. An all-too-frequent causal factor in accidents is "continued VFR flight into adverse weather." However, armed with knowledge of these phenomena, a pilot can avoid the hazard and nullify the risk.

The IFR pilot is also concerned with ceilings and visibility, especially in the takeoff, approach, and landing phases. Few general aviation aircraft are equipped with fully hands-off, auto landing systems. Therefore, low ceilings and visibility in takeoff and landing phases are critical to IFR operations. No segment of aviation is immune. A U.S. Air Force T-43 (Boeing 737) crashed in low ceiling and visibility conditions in Bosnia, killing all aboard, including U.S. Secretary of Commerce Ron Brown.

Of those pilots involved in low-ceiling accidents, more than half had instrument ratings. This should dispel the notion that instrument rated pilots are immune to low ceiling accidents. Also, more than half were fatal. In about one-third of the cases, there was no record of a weather briefing. A significant number of accidents occurred to IFR pilots descending below approach minimums.

In accidents involving fog, more than half were fatal. In more than two-thirds the weather was reported as IMC, and there was no record of a weather briefing. Again, IFR pilots are not immune. Descending below minimums and improper approach procedures were significant causal factors. Like fog, rain and other low-visibility accidents closely match.

In accidents involving snow, about one-third were fatal. However, rather than ceiling and visibility, the primary accident causal factor was loss of control on landing.

Although low ceilings and visibility are often found together, there are situations where only one or the other phenomena exist. Therefore, we have three situations: low ceiling with good visibility, low visibility with no significant ceiling, and the worst case—both low ceiling and visibility.

A *ceiling* is defined as a broken ($\frac{5}{8}$ to $\frac{7}{8}$ coverage) or overcast ($\frac{8}{8}$ coverage) layer. A ceiling restricts or precludes a VFR pilot from climbing or descending through the layer.

Ceiling heights are either measured or estimated. Reported ceilings may not be representative of the surrounding area. A sage axiom in aviation is "Weather reports may not be accurate, but they're official." This doesn't mean reports should be ignored; however, they should be viewed with caution, especially during marginal conditions and at night.

There are two types of ceilings. One measures a distinct cloud base. This is the first broken or overcast layer aloft in a METAR report. The second is an indefinite ceiling. An indefinite ceiling is reported when an obscuring phenomena (fog, snow, rain, smoke, etc.) hides the whole sky. It, in fact, is the vertical visibility into a surface base obscuration.

In the METAR code, restrictions to visibility caused by fog are reported using the weather phenomena codes FG (fog) or BR (mist). Why both FG and BR? The fact that these codes were developed by the British and French explains a lot. Allegedly, the British only consider it fog when you can cut it with a knife. Therefore, FG is reported when the visibility is less than $\frac{5}{8}$ of a mile. When fog restricts visibility between $\frac{5}{8}$ mile and 6 miles, it is reported as BR.

I'm sure everyone has noticed that the letters BR are not contained in the word mist. Why? The contraction BR, for mist, comes from the French word "brume." In METAR a dozen precipitation and weather phenomena contractions are derived from French words. You might also wonder who pulled $\frac{5}{8}$ mile out of the hat. It happens to correspond to 1000 meters. Another aviation mystery solved!

A partial obscuration is reported when between $\frac{1}{8}$ and $\frac{7}{8}$ of the sky is hidden by a surface-based obscuring phenomena. Precipitation, haze, smoke, and fog usually cause this condition.

Refer to Fig. 2-6. In the example, half, or $\frac{4}{8}$, of the sky is hidden by fog. The observer sees another $\frac{1}{8}$ cloud cover at 3000 feet. In the METAR code, a partial obscuration is indicated as FEW, SCT, or BKN on the surface (FEW000, between $\frac{1}{8}$ and $\frac{2}{8}$ of the sky obscured). In Fig. 2-6, our observer is reporting visibility reduced by fog (FG), between $\frac{3}{8}$ and $\frac{4}{8}$ of the sky obscured, with a ceiling of 3000 broken. (The observer used the summation principle of cloud cover; $\frac{4}{8}$ plus $\frac{1}{8}$ equals $\frac{5}{8}$. Five-eighths is reported as broken. The remark tells us that it is an obscuration (…RMK FG SCT000). What else could it be? Well, technically a layer with a base less than 50 feet would be reported as "000." This is very unlikely. If the observer did indeed mean to report a layer with a base less than 50 feet, the remark "FG SCT000" would not appear.

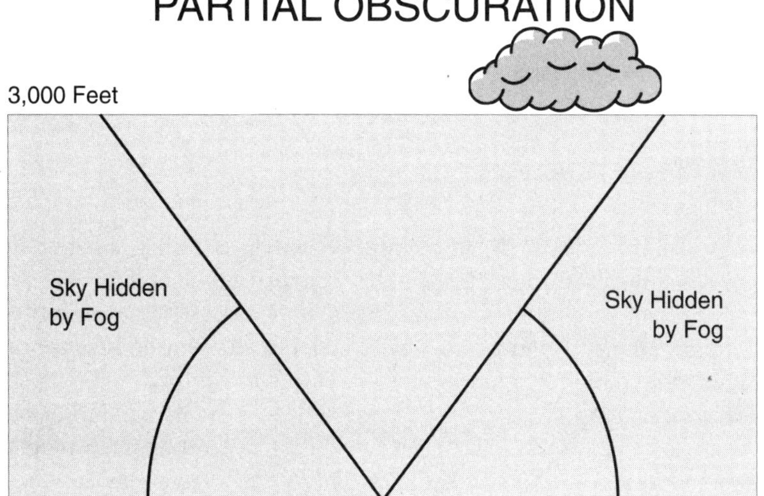

Fig. 2-6. *A partial obscuration is reported when the observer cannot evaluate all of the sky.*

Why is such a complicated procedure needed to report a partial obscuration? METAR has no provision to report this phenomena. The FAA and NWS concurred—probably for the first time—and proposed that a partial obscuration be eliminated from our reporting procedures, but there are other players in the game, namely the Department of Defense (DOD). (As of this writing, the FAA and NWS are still attempting to eliminate partial obscurations from reporting criteria.)

Normally, a pilot can expect ground contact while flying in areas with a reported partial obscuration. This is why they are not considered ceilings. Penetrating a partial obscuration VFR requires that appropriate horizontal visibility be maintained for that airspace.

After being briefed that the destination was reporting sky partially obscured, visibility 2 miles in mist and haze, a Baron pilot emphatically demanded to know the ceiling. The pilot stated this information was required to determine IFR minimums. There was no ceiling! IFR minimums, in this case, would be based solely on visibility. The pilot could expect to maintain ground contact throughout the approach, sighting the airport at about 2 miles.

Visibility is a measure of the transparency of the atmosphere. During the day, visibility represents the distance at which predominant objects can be seen; at night, visibility is the distance that unfocused lights of moderate intensity are visible. One National Weather Service (NWS) observer was quite perplexed when the tower always reported increased visibility after sunset. The reason was the change in criteria for the observation. Pilots should note that daytime values do not necessarily represent the distance that other aircraft can be seen. At night, especially under an overcast, unlighted objects might not be seen at all, and there might be no natural horizon.

Refer to Fig. 2-7. In aviation there are three distinct types of visibility:

- Surface
- Slant range
- Air to air

Surface visibility is visibility present at the surface. Prevailing visibility, reported in manual observations, is the greatest distance that can be seen throughout at least half the horizon circle, which need not be continuous. Automated stations determine visibility, normally only at one point on the airport using a backscatter device. In the absence of observations, pilots are required to determine surface and inflight visibility.

Slant range, or air to ground, visibility can be restricted by phenomena aloft or on the surface. It may be greater than or less than surface visibility, depending on the depth of the surface restriction and the intensity of the phenomena aloft. All pilots are concerned with slant range visibility on approach and landing.

Air to air, or inflight, visibility is visibility aloft. FARs require minimum inflight visibility for VFR operations to "see and avoid" other aircraft. These minimums increase with altitude and change with different classifications of airspace. Clouds and obscuring phenomena aloft often limit air to air visibility.

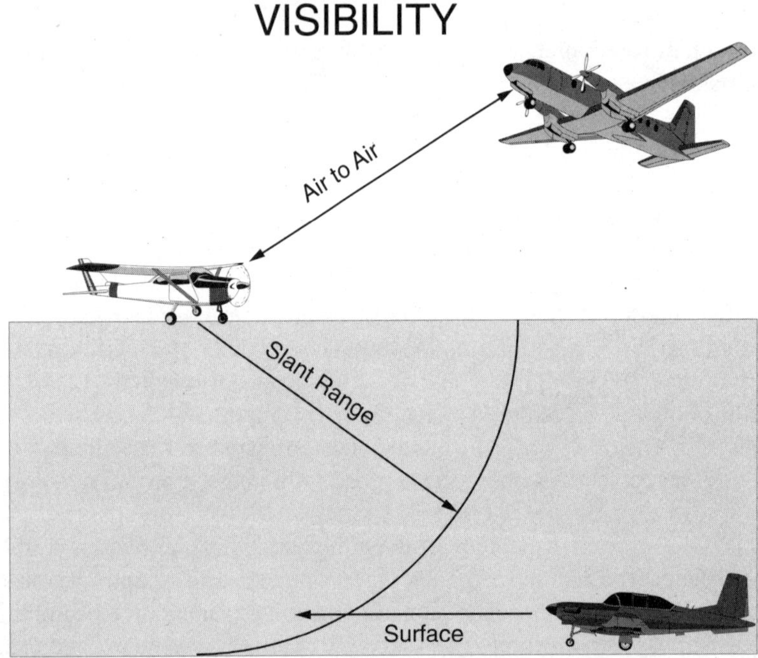

Fig. 2-7. *In aviation there are three distinct types of visibility: surface, slant range, and air to air.*

Visibilities aloft are most often reduced by rain, snow, dust, smoke, and haze. And, with reduced visibilities, usually less than 5 miles, the apparent visibility looking toward the sun can be almost nil! Pilot reports are the only source of slant range and inflight visibility.

Obstructions to vision caused by fog, haze, dust, and smoke are reported in weather observations when visibility is less than 7 miles. When these phenomena exist with visibilities 7 miles or greater, a remark might describe the condition. Pilot weather briefers sometimes use the term unrestricted to describe visibilities of 7 miles or greater. This sometimes causes confusion. A pilot told visibility unrestricted might respond, "What about the haze?" Thus, the phrase *visibility unrestricted* does not imply that smoke, haze, dust, or even fog are not present, just that visibility is 7 miles or greater.

Areas of low ceilings and visibilities are graphically depicted on weather depiction charts, advertised in the AIRMET bulletin, reported in METAR reports, and forecast in TAFs.

Fog

Fog, a cloud with its base on the ground, occurs most frequently in coastal regions due to the large amount of water vapor available. However, fog can form anywhere. The rapidity with which fog can form makes it especially hazardous. It is not unusual for visibility to drop from more than 3 miles to less than half a mile in a few minutes.

Fog forms by any atmospheric process that does one or both of the following:

- Cools the air to its dewpoint
- Raises the dewpoint to the air temperature, adding water vapor

Fog, ground fog, and ice fog describe the same condition. *Ground fog,* normally less than 20 feet deep, reduces visibility horizontally rather than obscuring the sky. Usually localized, formed by radiational cooling, ground fog tends to dissipate rapidly once clearing begins. *Ice fog* forms in cold weather at temperatures around −30°C from radiational cooling and exhibits the same characteristics as radiation fog.

Radiation fog forms when air cools from contact with the ground and becomes saturated. This occurs at night and tends to be most dense, with lowest visibilities, around sunrise. Clear skies, light winds (less than 5 knots), high relative humidity, and stable air are favorable conditions for the formation of radiation fog. We have already discussed how low water vapor content in the atmosphere increases terrestrial radiational and cooling; therefore, dry air aloft enhances the formation of radiation fog. Radiation fog shows up well on visual satellite imagery. Overcast skies, strong winds, low relative humidity, and unstable air prevent or retard the formation of radiation fog.

Radiation fog tends to be patchy and shallow, usually burning off by mid-morning. It tends to form in valleys after moisture has been added at the surface from passing storms. As high pressure—clear, stable conditions—build into an area, circumstances are right for the formation of radiation fog. This condition can become persistent in California's Central Valley during winter and early spring. (In this area, the low fog is known as "Tule fog" with tops usually less than 3000 feet. "Tule" [tôô′lê] is a Spanish word for bulrushes, a marsh plant that grows during this season.) Figure 2-8 shows Tule fog in California's Central Valley. Note how well the lateral extent of the fog shows up on satellite imagery. The satellite clearly indicates that the fog extends from the Central Valley through the Carquinez Straits into the San Francisco Bay.

Other areas conducive to the development of radiation fog are the valleys east of the Cascade Mountains in the Pacific Northwest, Great Basin, Snake River Valley, and valleys of the Appalachian Mountains.

Radiation fog does not usually form until the second day after storm passage due to atmospheric mixing following the front. A strong inversion develops, which locks moisture at lower levels, and radiation fog forms. Zero-zero conditions over widespread areas can persist for days or even weeks until the moisture evaporates or another storm system moves through the area.

IFR pilots normally have no difficulty operating in conditions caused by radiation fog, as long as they don't mind flying above zero-zero surface conditions. And, landing minimums might not prevail until late morning or afternoon, if at all, especially in valleys with fog depths of 2000 to 3000 feet. Why does this occur? During the winter months days are short, reducing the length of time for burnout (evaporation) by the sun. Longer nights provide longer periods for the fog to develop. The VFR pilot will be delayed until the condition dissipates. Some pilots routinely move their aircraft to mountain or higher-elevation airports above the fog layer during winter months.

Fig. 2-8. *Satellite imagery is very effective for determining the extent of fog, as long as higher clouds are not present.*

Expect radiation fog the following morning when at dusk skies are clear, wind is light, and temperature–dewpoint spread is 8°C or less. Since fog evaporates from the edges, radiation fog tends to dissipate from the mountains toward the center of the valley. Unfortunately, many airports are located at the lowest elevations in the center of the valley, where fog remains the longest.

Advection fog forms when moist air moves over colder ground or water. The air cooled from below becomes saturated. Unlike radiation fog, advection fog can form under an overcast. This is a persistent condition along the Pacific coast during the summer months. The prevailing onshore flow moves the layer into coastal sections and valleys. It is usually deepest and farthest inland at sunrise and retreats toward the ocean during the day. Figure 2-9 illustrates the effectiveness of visual satellite imagery for determining the extent of the layer as long as higher cloud layers are not present. It is especially useful in areas without weather reporting stations. Winds of 5 to 15 knots tend to cause low ceilings rather than fog.

One feature of the coastal stratus is the eddying effect of coastal winds, in particular the "Catalina eddy." The Catalina eddy is a southeasterly current of air along the immediate coast of Southern California flowing contrary to the main northwesterly flow. Its vortex is in the vicinity of Catalina Island. This southeasterly current disappears above the temperature inversion, which caps the layer of cool, moist air in which the eddy is embedded. It usually develops when a rather strong current of air flows southeastward over the ocean near the coast with falling pressures inland. Coastal stratus increases rapidly, often unforecast, and is carried further inland with bases as well as tops higher than normal. A solid overcast well inland may persist for several days.

Upslope fog forms as air is forced upward, expands, and cools adiabatically. Moist air must be forced upslope, which requires a wind of 5 to 15 knots. This condition occurs during winter and spring in the Midwest, where terrain rises steadily from the Gulf of Mexico to the Rockies. Figure 2-10 shows an area of extensive upslope fog that has developed over the Texas panhandle.

Upslope fog can be widespread and persists as long as favorable conditions continue. During upslope fog conditions, the VFR pilot is pretty much out of luck. The IFR pilot might not be much better off. He or she might encounter IFR landing minimums, but the condition often exists over areas the size of several states. A legal IFR alternate might be beyond the aircraft's range. Here again, satellite imagery is useful in determining the extent of upslope fog, as long as higher cloud layers are not present.

Expect upslope fog and low clouds when the wind blows up sloping terrain and the temperature–dewpoint spread is small enough that air cools to saturation.

Rain-induced fog, also known as frontal fog, occurs when warm rain falls through cooler air, evaporates, and condenses, forming fog; it can be dense and persists as long as the rain continues. Winds must generally be light. This condition is usually associated with stationary, warm, shallow, or weak cold fronts. A potential for clear icing exists in areas of rain-induced fog. Satellite imagery is of no use in determining the extent of the fog because of the presence of higher cloud layers.

Expect rain-induced fog when the temperature–dewpoint spread is small and continuous rain or drizzle is falling in areas associated with stationary, warm, and weak cold fronts.

Fig. 2-9. *Like radiation fog, the extent of advection fog is clearly depicted on visual satellite imagery.*

Steam fog develops as cold air moves over warm water. Evaporation from the water takes place and saturation occurs. Low-level turbulence develops as the warm water heats lower levels, creating a shallow layer of instability. Also known as *evaporation fog,* steam fog occurs in cold climates over lakes, such as the Great Salt Lake and the Great Lakes in the autumn. Expect steam fog when the wind is blowing from a cold land area over warm water.

Look for the development of fog when wind is blowing from a warm surface over a cold surface and dewpoint is warmer than the colder surface. In general, be prepared for fog when the temperature–dewpoint spread is 3°C or less and decreasing. But remember it usually takes more than just a close temperature–dewpoint spread for fog to develop. Expect little if any improvement in visibility when fog exists below overcast skies or when rain or drizzle is forecast to continue.

At times, when conditions are favorable for fog, a very low cloud layer forms. This is especially true over flat terrain when winds exceed about 15 knots. These fog-like clouds, formed in stable air with smooth flying conditions, often exist together with fog.

Haze and Smoke

Haze is caused by the suspension of extremely small, dry particles invisible to the eye but sufficiently numerous to reduce visibility. Haze, combined with smoke, often describes conditions in metropolitan areas—sometimes translated as hack and kough. Large anticyclones—high pressure cells—can dominate the southeastern United States, trapping haze and pollutants, especially in industrial areas. Above the haze layer, visibilities are unrestricted and temperatures cool, resulting in a much more comfortable flight.

Surface inversions form at night due to radiational cooling. As the surface cools, the lower layer of the atmosphere cools while temperatures aloft remain unchanged. Inversions aloft often trap haze and smoke, sometimes reducing visibility aloft to less than 3 miles.

Strong inversions over cities or industrial areas can trap haze and smoke, reducing visibility to less than 1 mile. Landmarks are all but invisible. At altitude, dense haze often appears solid, like a cloud layer, in the distance; this occurs with high relative humidity. When the sun strikes the layer, light waves scatter, causing the layer to appear white. This accounts for apparent inconsistencies in surface observations. Reports might report clear or partially obscured skies with visibilities 3 to 5 miles in haze and smoke. A pilot looking into the sun often has no forward visibility or natural horizon. Pilots caught in such situations, technically VFR, have become spatially disoriented and lost control of their aircraft.

Fig. 2-10. *Upslope fog can be widespread and persists as long as favorable conditions continue.*

Figure 2-11 illustrates a typical hazy day, under mostly clear skies, and a stable air mass. Notice how the terrain becomes obscure in the background and even appears as if it might be clouds. The light band in the middle of the picture is an area of relatively clear air above the surface-based haze layer and below another haze layer aloft.

Haze can obscure mountain ranges in the West, making pilotage navigation difficult. Under hazy conditions, it's all but impossible to distinguish one range from another. In the Midwest and East, haze can obscure thunderstorms.

Haze and smoke are usually restricted to an area below 5000 feet, although layers can extend to above 10,000 feet. During the devastating September 1987 forest fires in California, smoke tops were reported as high as 19,000 feet, with visibilities aloft zero.

At Redding, California, an NWS observer reported an indefinite ceiling 3000 feet; the sky was completely hidden by smoke and haze. A pilot flying in this area can only expect to maintain ground contact within about 3000 feet of the surface. This situation could be extremely dangerous for the VFR pilot. Should the pilot climb above 3000, he or she would be in IFR conditions without ground contact and most probably without the natural horizon. Pilots attempting to operate in similar conditions have lost aircraft control with fatal results. Figure 2-12 is a satellite image of these conditions. The light gray areas of Northern California, Oregon, and eastern Washington are smoke.

Expect restrictions to visibility due to haze and smoke in areas of stable air and light winds. Visibilities may be less than 1 mile with strong, low inversions. Surface heating

Fig. 2-11. *Haze trapped in inversions aloft can reduce visibility at altitude as well as on the surface.*

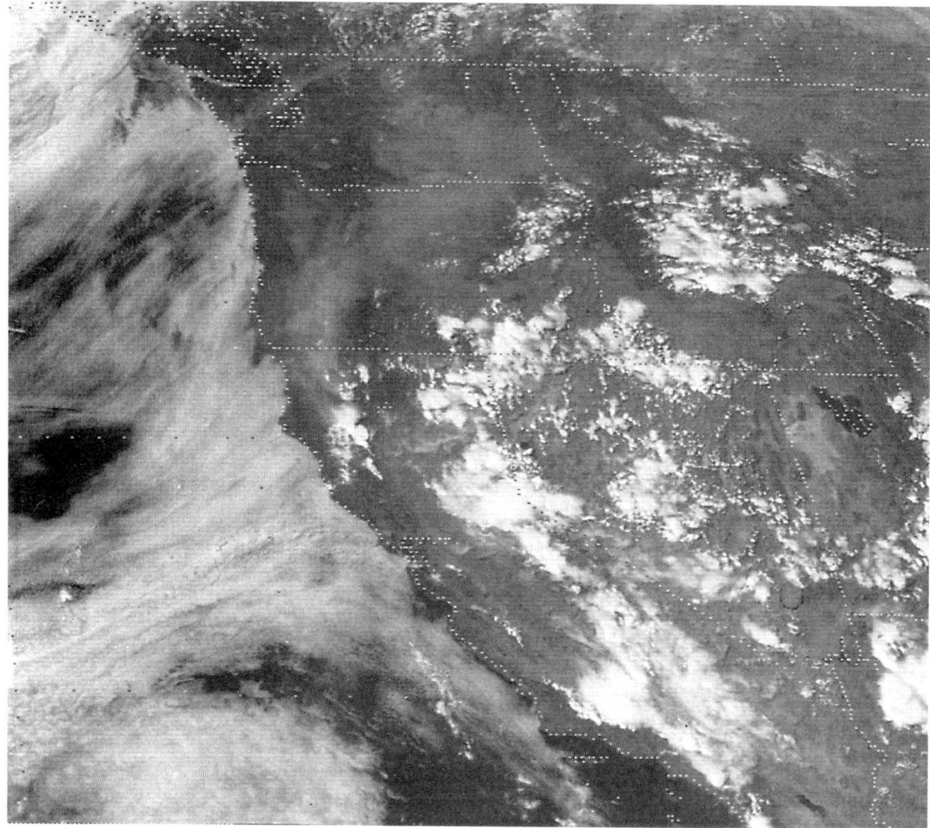

Fig. 2-12. *Extensive smoke layers are visible on this satellite image; the light gray areas of Northern California, Oregon, and eastern Washington are smoke.*

often breaks through the inversion, and along with increased winds, visibilities tend to improve during the afternoon. This is a typical case during the summer and fall in the Los Angeles Basin. Expect little if any improvement in visibility when haze or smoke exist under an overcast.

The following observation was taken at March AFB, California: KRIV…RMK CREPUSCULAR RAYS SW. *The Glossary of Meteorology* defines *crepuscular rays* as "literally, 'twilight rays'; alternating lighter and darker bands (rays and shadows) that appear to diverge in fan-like array from the sun's position at about twilight." Towering cumulus clouds produce this effect, especially with haze in the lower atmosphere. This seems a rather complicated way of saying: HAZY TCU SW (hazy with towering cumulus southwest).

Dust, Sand, and Volcanic Ash

Dust and blowing dust, a combination of fine dust or sand particles suspended in the air, can be raised to above 16,000 feet by the wind. Visibilities, surface and aloft, can be at or near zero. Because of its fine particles, dust can remain suspended for days once the wind subsides. Figure 2-13 shows blowing dust being raised from the surface in California's Mojave Desert by strong winds.

Blowing sand, made up of particles larger than dust, usually remains within a few hundred feet of the surface. It can also reduce visibility to near zero. But when the wind subsides, particles fall back to the surface and visibility improves rapidly. A pilot approaching Lovelock, Nevada, skeptical of a reported visibility of 2 miles in blowing sand, reported his flight visibility was 20 miles. Upon landing, however, he concurred, stating the tops of the blowing sand were at 200 feet above ground level (agl).

When blowing sand reduces visibility to less than 3 miles, the condition is advertised in a SIGMET (significant meteorological information). Expect reduced visibilities due to blowing dust and sand when strong winds are forecast and terrain surface is barren. This is especially true in deserts and the Great Plains.

This observation came from an NWS observer at Denver, Colorado: KDEN...RMK DSIPTG GUSTNADO N (dissipating gustnado north—a glorified dust devil). *Gustnado* is a local term used to describe a funnel cloud that develops along the gust front of a thunderstorm. It is believed that the gustnado receives its initial rotation from the shift in wind directions across the gust front. Cold, dense air behind the gust front lifting the warm air ahead imparts a rotating motion in the windshear zone.

Fig. 2-13. *Because of its fine particles, dust can remain suspended for days once the wind subsides.*

Dust devils, reported as dust or sand whirls in the METAR code, are whirlwinds that form on clear, hot days with light winds. They have diameters of 10 to 50 feet and extend from the surface to several thousand feet. Windspeeds within the rotation vary from 25 to more than 75 knots. Dust devils are capable of substantial damage, but the majority are small. We flew into a dust devil doing pattern work at Lancaster's Fox Field in California's Mojave Desert. The encounter was equivalent to light to moderate turbulence, it shook the Cessna 150, and the low pressure in the vortex caused both windows to pop open!

Volcanic ash became a significant weather phenomena in the continental United States with the eruption of Mt. St. Helens in 1980. A typical eruption can release 500 million tons of ash into the atmosphere. Since 1980 there have been a hundred or so eruptions worldwide, and more than a dozen in the United States—mostly in Alaska. Volcanic ash is difficult to distinguish from clouds. Volcanic ash is seldom detected on ATC or aircraft radar. However, NEXRAD, the NWS's advanced Doppler radars, where available, should provide a new tracking capability for volcanic ash.

Volcanic ash consists of fine particles of rock powder, blown out from the volcano. The particles remain suspended in the atmosphere for long periods, extend well into the flight levels, and may drift thousands of miles. Ash power, up to ⅛ inch in diameter, can be very abrasive. Volcanic ash can be extremely destructive to aircraft leading edges, windscreens, and engines. Turbojet aircraft engines are especially susceptible.

Royal Dutch Airlines Flight KLM867, a Boeing 747, encountered an ash cloud from the eruption of Mt. Redoubt in 1989. The aircraft lost all four engines to flameout! ATC vectored the aircraft back to Anchorage. During the descent the crew was able to restart two of the engines and make a safe emergency landing. KLM867 was one of several aircraft damaged by the eruption. Such encounters happen every year.

Ash on the airport is another significant hazard. Ash can be blown into the air, causing reduced visibilities and aircraft damage. Wet ash on the runway reduces breaking action, and crews are trained to reduce or avoid the use of reverse thrust.

Volcanic ash is also a hazard to general aviation aircraft. Ash affects windscreens and leading edges. It blocks air filters and pitot systems. We have already discussed the hazard to turbojet engines—many of which are used on corporate twins and helicopters. And the hazards of reduced visibilities and slippery runways cannot be overlooked.

Like icing and thunderstorms, the key to dealing with volcanic ash hazard is avoidance. Do not fly into or climb through ash clouds. If inadvertently entered, reverse course and descend. Finally, report the encounter to alert other pilots of the hazard and assist officials to track the cloud. Reports and forecasts of volcanic ash are available in METAR reports, SIGMETs, Center weather advisories, and the Volcanic Ash Forecast Transport and Dispersion Chart.

Precipitation

Snow, drizzle, and rain are the most common forms of precipitation restricting visibility. Of these, snow is usually the most effective in reducing visibility. Heavy snow frequently reduces visibility to near zero. On the other hand, rain rarely reduces visibility to below 1 mile and has a tendency to wash haze and smoke out of the air. Cold rain even removes fog from the air. Conversely, drizzle often accompanies fog, haze, and smoke, resulting in lower visibility than occurs in rain.

Figure 2-14 shows a downburst over the mountains west of Reno, Nevada. Notice how the precipitation obscures the terrain, another hazard associated with this phenomena. The AIRMET Bulletin warns pilots about mountain obscuration in clouds and precipitation.

Rain or drizzle on the windscreen can reduce the pilot's visibility considerably below visibility outside the aircraft. Dry snow does not adhere to the windscreen and does not greatly affect visibility through the windscreen. Blowing snow, like blowing sand, reduces visibility near the surface when strong winds blow over freshly fallen snow. Visibility can be near zero close to the surface, with rapid clearing after the wind subsides. Expect reduced visibility due to blowing snow when strong winds are forecast and terrain surface is snow-covered.

Whiteout is an atmospheric optical phenomenon in which the pilot appears to be engulfed in a uniformly white glow. Neither shadows, horizon, nor clouds are discernible; sense of depth and orientation is lost. Whiteout occurs over an unbroken snow cover and beneath a uniformly overcast sky when light from the sky is about equal to that from the snow surface. Blowing snow may be an additional cause.

To the VFR pilot, whiteout is disastrous. Snow-covered terrain, an overcast, and already reduced visibility are a strong no-go indicator. At the very first sign the pilot's only option is 180° turn to what, hopefully, were better conditions.

Fig. 2-14. *Precipitation obscures the terrain, another hazard associated with this phenomena.*

The IFR pilot is not immune to whiteout. In one instance, a pilot's first destination did not have an instrument approach. It was snowing and the pilot reported whiteout conditions. The pilot diverted to an airport with an instrument approach.

Reported weather was visibility ½ variable between ¼ and 1 mile in snow, indefinite ceiling 600 feet. Radio contact was lost after the second missed approach. The wreckage was located an hour later.

The following is strictly speculation. Because the crash occurred after the declaration to miss the approach, it appears the pilot decided to miss after the missed approach point. The transition between instrument and visual flight under these conditions is extremely difficult, especially with a single-pilot operation. The pilot may have acquired ground contact straight down, but slant range and apparent whiteout conditions, would preclude visual contact with the approach environment and airport.

Many commercial operations require two pilots for just this scenario. One pilot stays on the instruments, and the other looks for the airport. I had a similar incident while training an instrument pilot. We were flying at night in rain and fog, and ceiling and visibility were at minimums. At the missed approach point, my student started the missed approach. At that point I caught a glimpse of the approach lights. We were certainly not in a position to land and continued the missed approach procedure.

Terrain Obscurement

We have already touched on the subject of terrain or mountain obscurement. Haze and smoke typically obscure distant landmarks, making visual navigation difficult. Precipitation, especially when heavy or accompanied by low visibility, can obscure terrain, as illustrated in Fig. 2-14. More typically, terrain is obscured by clouds.

To alert pilots to the potential dangers of cloud or precipitation obscurement, the NWS advertises these conditions through a weather advisory. The AIRMET Bulletin contains a section, AIRMET SIERRA, addressing mountain obscuration. This forecast outlines areas where extensive or widespread obscurement is expected.

It might be a good idea to describe the difference between IFR conditions—which are also advertised in AIRMET SIERRA—and mountain obscuration. (But, before we begin, we must make an important point. To be included in a weather advisory, the phenomena must be widespread. Local or isolated conditions are not within the scope of these products. More about this later.) Typically, advisories for IFR conditions apply to flat, nonmountainous terrain, for example, coastal valleys, California's Central Valley, the high plateau of the intermountain region of the West, and the Great Plains. Advisories for mountain obscurement apply to the mountains—coastal mountains, Cascades, Sierra Nevada, Rockies, Appalachian, and so on.

Mountain obscurement infers that VFR flight through valleys is usually possible, but visual flight over mountains or through mountain passes may not be possible. Figure 2-15 illustrates mountain obscurement. With good visibility, this hazard should not present a problem. Notice in Fig. 2-15 that VFR conditions are good in the valleys around the mountain, with the peak hidden in clouds. These clouds are sometimes classified as "cumulo-granite."

Fig. 2-15. *Clouds associated with mountain obscurement are sometimes classified as "cumulogranite."*

Returning from Oshkosh, Wisconsin, we planned a leg from Pueblo, Colorado, to Farmington, New Mexico. Low pressure was affecting the weather, and an advisory for mountain obscurement was in effect. We successfully negotiated the Sangre de Cristo range between Pueblo and Alamosa, Colorado. VFR flight was no problem in the San Luis Valley. However, the San Juan range between Alamosa and Farmington was not to be conquered. We tried to top the clouds without success. Plan B was to fly south toward Taos, where terrain was lower. West of Taos a safe pass allowed us to proceed uneventfully—except the increased flight time made that last cup of coffee at Pueblo most uncomfortable. We always had an out. If the pass was closed, both Taos and Alamosa could be used as alternates. (One more comment from a practical point of view. Throughout the flight, we updated weather, our route of flight, and estimated time of arrival with Flight Service.)

STRATEGIES

The pilot departed a Southern California airport during the evening for a flight to Monterey. Arriving in the Monterey area about midnight, the pilot found the airport overcast. The pilot landed on a highway south of the Reid–Hillview airport in the San Francisco Bay area. There was no record of the pilot checking en route for a weather update, which would have revealed the onset of coastal stratus. There is no rational reason for this incident to have occurred. The pilot landed safely and was applauded by some.

Unfortunately for aviation, opponents of Reid–Hillview cited it as another reason to close the airport.

In the preceding example, even the FAR-required fuel reserve may not have been adequate. A 45-minute fuel reserve doesn't make any sense with the nearest suitable alternate 50 minutes away. What would have happened if the coastal stratus extended beyond the airplane's range? Most likely the accident would have been fatal. How could this incident have been prevented? Simple—update weather en route with Flight Watch or Flight Service.

When conditions are favorable for coastal stratus, VFR pilots should plan arrivals and departures during the afternoon hours. If this is not possible, moving the aircraft to an airport a few miles inland often allows a morning departure.

It's often hazy in the southern portion of California's San Joaquin Valley. A pilot flew for more than an hour looking for the Porterville airport. The pilot ran out of fuel and landed in a field. In spite of talking to unicom, this pilot could not locate the airport. A week or so later another pilot had a similar experience under the same weather conditions. Unable to locate the Porterville airport, this pilot called air traffic control. Specialists at the flight service station provided a direction finder (DF) steer to Bakersfield. The FAA has personnel and equipment ready and waiting, but the pilot has to ask for assistance. Let ATC help before an incident becomes an accident.

Every year, pilots become lost, even lose control of the aircraft, flying in reduced visibility. Often flight conditions improve by climbing to a higher altitude. Once above the layer, slant-range visibility is usually greater with a distinct horizon preventing disorientation. The seemingly obvious assumption the closer to the ground the better to see it—usually isn't true.

One pilot was operating in the southeastern United States, in an area of reduced visibility in haze and fog. The noninstrument-rated pilot had difficulty maintaining heading and altitude. As it was approaching the airport to land, the aircraft was seen in a descending right turn. The pilot crashed into trees and hit the ground. The crash was fatal. Visibility in the vicinity of the crash site was estimated to be between 1 and 2 miles in fog. The pilot told several others of concern about flying in hazy conditions.

This pilot's first error was flying in conditions below FAR weather minimums. The pilot was obviously uncomfortable about the flight. If that's the case, why depart in the first place? What can be so important as to endanger your life? If any doubt exists, don't go! Second, the pilot could have climbed to a higher altitude, above the haze and fog layer, which may have allowed the pilot to maintain aircraft control. Once on top, the pilot could have obtained the services of ATC to locate or assist in finding a suitable landing site.

With low ceilings and visibility, the greatest hazard occurs when both prevail. We were on a flight from Shreveport, Louisiana, to Mineral Wells, Texas. Ceilings were below 1000 feet, but visibility was unlimited. In the sparsely populated Texas countryside, it was easy to maintain legal distance from objects on the ground and avoid towers and power lines. As we flew south of Dallas, the clouds lowered to the ground. In the noninstrument-equipped Cessna 150, there was only one course of action—reverse course and land. We were delayed another day before we could proceed west.

Pilots with even limited instrument capability—I'm speaking of both aircraft and equipment—have the option of filing IFR. Always the best situation is to "get into the system" before reaching IFR conditions. Here are some examples.

When I first came into the FAA, I was assigned to the Lovelock, Nevada, FSS. I would routinely fly from Lovelock to Van Nuys, California. My Cessna 150, although equipped and certified for IFR, did not have the capability to fly IFR over the Sierra Nevada mountains. But if the only weather was coastal stratus, I would file VFR to Palmdale and IFR from Palmdale to Van Nuys. It's always easier to pick up a profiled clearance before reaching congested terminal airspace.

On one of our trips to Oshkosh, the leg from Mason City, Iowa to Madison, Wisconsin, was plagued with marginal VFR weather toward the destination. Rather than try to fly under the weather or attempt to pick up IFR in the Madison area, I elected to file IFR from Mason City. The weather was clear, so we departed VFR and picked up the clearance with Center on climbout. It was a good plan. With marginal weather in the Madison area, controllers were refusing pop-ups and instructing VFR aircraft to remain clear of Class C airspace. Pilots were advised to contact the next sector for traffic advisories. There is only so much airspace, especially in terminal areas.

Plan A doesn't always work. After remaining overnight at North Platte, Nebraska, we planned to continue westbound to Cheyenne. A front had passed through the previous day and the ground was moist. The airport, of course, is next to the river, where the fog and low clouds are usually the worst. The weather to the west was good, and I obtained a special VFR clearance. After takeoff it became apparent it wasn't going to work. We had no option except to land, return to the airport office, and file an IFR flight plan. Twenty minutes later we were on our way. (When you have time to spare, go by air.)

An AIRMET was in effect for mountain obscurement in Idaho. During the briefing for the flight from Winnemucca, Nevada, to Idaho Falls, the briefer advised that "V-F-R flight was not recommended" (VNR). This was a little silly because our flight took us through the Snake River Valley, not the mountainous areas that were obscured. Upon opening our flight plan and again updating weather en route, we were VNR'd. The valley was perfectly fine.

The next morning we prepared to fly from Idaho Falls to Billings, Montana. Weather at both ends was good, with no weather advisories for the route. What we found over Yellowstone National Park is illustrated in Fig. 2-16. As we approached the park, the valleys were fogged in, and there were several layers aloft. Most of the mountains were obscured. Because of the minimum en route altitudes and freezing levels, IFR (on purpose) was not an option. We picked our way through, at times climbing to more than 13,000 feet in the Cessna 172. We always had the option to reverse course and return to Idaho Falls or divert into West Yellowstone. We were lucky, and on the north side of the park, we were able to descend below the clouds, verify with Flight Watch that Billings weather was good, and proceed on to our destination.

This scenario has the potential for disaster. Navigation was difficult because of sparse navigational aids and terrain obscurement. As is my practice, I had a dead reckoning course planned and calculated. Throughout the flight I was evaluating options based on weather and fuel: proceed, return, or divert.

Fig. 2-16. *Pilots must use caution when flying between cloud layers; they aren't called "sucker holes" for nothing!*

Oh, a side note. We were not VNR'd on the flight to Billings. There were no weather advisories. The point is that the existence, or lack, of an advisory does not preclude, or guarantee, a safe flight. Pilots must evaluate each flight separately, based on the reported and forecast weather, and their aircraft, training, and experience.

Recall that we mentioned that no segment of aviation is immune from low ceiling and visibility accidents. The typical IFR scenario is descending below minimums. It has happened to the airlines and the military, as well as general aviation pilots. We mentioned the Air Force T-43 crash in Bosnia. Investigators discovered, among other things, that the aircraft was not properly equipped for the approach to be used and was not flown at proper airspeeds. It's foolhardy to attempt an approach without the appropriate equipment functioning normally. This includes appropriate and current charts. Anytime during the approach if anything isn't normal, abandon the approach!

The dos and don'ts of flying in low ceilings and visibility:

- Do obtain a complete weather briefing.
- Do update weather en route.
- Do get into the IFR system before encountering poor weather.
- Do ask for help whenever the situation becomes doubtful or uncertain.
- Do have options and a plan for each option.
- Don't let the briefer make the decision (VNR or no VNR).

- Don't fly below legal or personal minimums.
- Don't run out of options. Land before you do.

If used, these strategies can prevent low ceiling and visibility accidents. Taking everything into account, all are preventable. Almost one-third of the pilots involved in low-ceiling and two-thirds involved in low-visibility accidents had no record of a weather briefing, and, although there are no specific records, it's doubtful they obtained weather en route.

3
Motion of the air

WITH A SOUND UNDERSTANDING OF THE ATMOSPHERE, ITS PROPERTIES, stability, and vertical motion we move on to movement of the air. Wind transports atmospheric properties in a process known as *advection.* As the air moves, its characteristics are influenced by the temperature and moisture properties of the surface; the atmosphere can be modified in the same way. In essence, the atmosphere is a giant heat exchanger, moving warm air up from the equator and cold air down from the poles. As we shall see, this movement has a tremendous effect on our weather.

We begin with planetary scale, or global, circulation, dealing with what is usually called *general circulation.* It consists of the jet stream, subtropical high, polar front, and intertropical convergence zone. We discuss the process by which the atmosphere transports warm air from the equator northward and the cold air from poles southward. These winds are caused by pressure differences in the atmosphere.

Synoptic-scale (*mesoscale*) circulation consists of highs and lows and troughs and ridges, both surface and aloft. Additional synoptic-scale events are thermal advection, fronts, and terrain effects. Highs and lows and troughs and ridges are discussed in Chapter 5. Fronts are presented in Chapter 6. Terrain effects are covered in this chapter.

From synoptic scale motion we move on to *subsynoptic-scale* (*microscale*) events; that is, sea/land breezes, mountain/valley winds, mountain waves, and windshear. Density, as well as pressure, plays a role in some of these events. In the case of mountain waves, terrain and atmospheric stability are key elements.

Finally, we touch on nonconvective low-level windshear, a microscale event; that is, low-level windshear caused by phenomena other than thunderstorms. (Because of the significance of thunderstorm hazard, including convective low-level windshear, they are covered separately in Chapter 5.)

Turbulence is a fact of life to the aviator. It ranges from small annoying bumps to a severity that can literally tear an aircraft apart. Turbulence can be on the mesoscale, covering areas the size of several states, or on the microscale, affecting one airport and leaving adjacent fields untouched. With one exception, turbulence is caused by natural phenomena. (OK, except in the case of poor pilot technique.) Wake turbulence is caused by the aircraft producing lift. Because of the hazard caused by wake turbulence, we decided to include it in our discussion of turbulence.

On the surface it may seem that pilots are doomed to live with turbulence (that's an oxymoron, isn't it?). This is not always true. There are strategies to eliminate or reduce the effects of this phenomena. The final section of this chapter provides strategies to deal with turbulence.

GENERAL CIRCULATION

Three-dimensional motion in the atmosphere results from differential heating of the earth's surface, which in turn causes pressure differences. In 1735, George Hadley added an important notion to the understanding of the general circulation. He proposed that the earth's rotation produced a deflective force on large-scale wind flows. Hadley was the first to propose a direct thermally driven and zonally symmetric circulation as an explanation for the trade winds. This circulation consists of the equator-ward movement of the trade winds between about 30° latitude and the equator in each hemisphere, with rising wind components near the equator, poleward flow aloft, and finally descending components at about 30° again. The *Hadley cell* is a direct thermally driven vertical motion system. There are typically three in each hemisphere, as illustrated in Fig. 3-1. The cells cause zonal wind patterns with strong easterly and westerly components, which cause the basic motion system of the general circulation.

The converging northeast and southeast trades produce the *intertropical convergence zone* (ITCZ). The ITCZ fluctuates in position and intensity; it can be weak and discontinuous, as shown in Fig. 3-2, a satellite view of the North Pacific Ocean. The source of the trades are the subtropical highs. Note the location of the high pressure cells ("H") in Fig. 3-2. It should be no surprise that in the vicinity of the ITCZ there is rising air (convergence), and, in the area of the subtropical highs, descending air (divergence).

Moving north, about 30° latitude, are the prevailing westerlies. These winds are also due to the subtropical highs. At the poles is another area of subsidence. This is the region of the polar high—some of the highest atmospheric pressures ever recorded have occurred in these areas. (In the late 1980s, pressures well above 31.00 in. Hg. occurred in Alaska. This precipitated emergency regulations because, as you recall, most altimeters can only be corrected for pressure up to 31.00 in., as mentioned in Chapter 1.) This region is also the area of the polar easterlies at about 60° latitude.

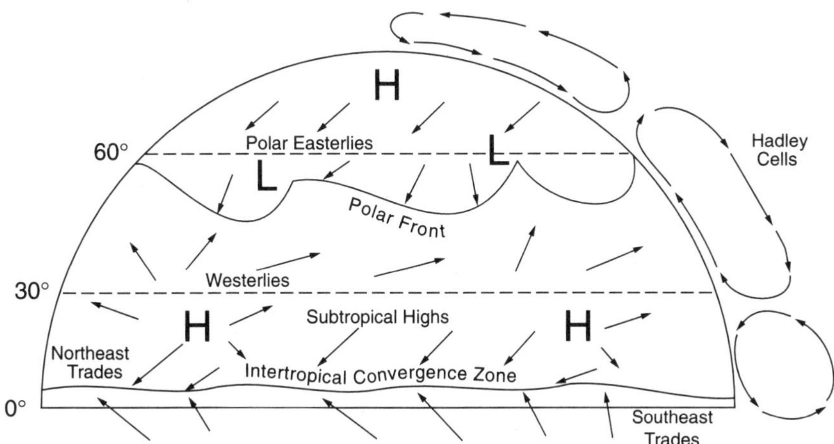

Fig. 3-1. *The general circulation consists of the jet stream, subtropical high, polar front, and intertropical convergence zone.*

Between the polar easterlies and the midlatitude westerlies is the polar front. The polar front is another area of global convergence of warm air from the south and cold air from the north. The polar front is more or less continuous around the world, as shown in Fig. 3-2. However, where it is weak, there may be areas of little or no weather over the continental United States, as illustrated in Fig. 3-2. Wind flow at the surface is depicted on surface analysis charts.

Wind is directed by three forces: pressure gradient force, Coriolis force, and frictional force. Differences in pressure cause *pressure gradient force.* At least initially, the wind wants to blow from areas of high pressure to low pressure. On weather maps, lines of equal pressure are called *isobars.* Isobars typically are nonconcentric circles surrounding centers of high and low pressure. If pressure gradient were the only force acting on the wind, wind would always blow perpendicular to isobars—directly from areas of high pressure to low pressure. However, this is not the case.

Pressure gradient force determines the strength of the wind. The stronger the pressure gradient force, the stronger the wind. On the surface analysis chart, areas of strong winds are identified by close spacing of isobars; areas of weak winds, by wide spacing of isobars.

Because of the earth's rotation, there is an apparent force that deflects the wind from a straight path relative to the earth's surface. This is *Coriolis force.* In 1856 American William Ferrel published "Essay on the Winds and Currents of the Ocean." He showed mathematically that winds are affected by the rotation of the earth. Gaspard de Coriolis, a French mathematician, in 1835 had already developed the theory of an apparent deflection force produced by angular rotation. Although Ferrel applied the theory to the winds, in meteorology we refer to this effect as Coriolis force.

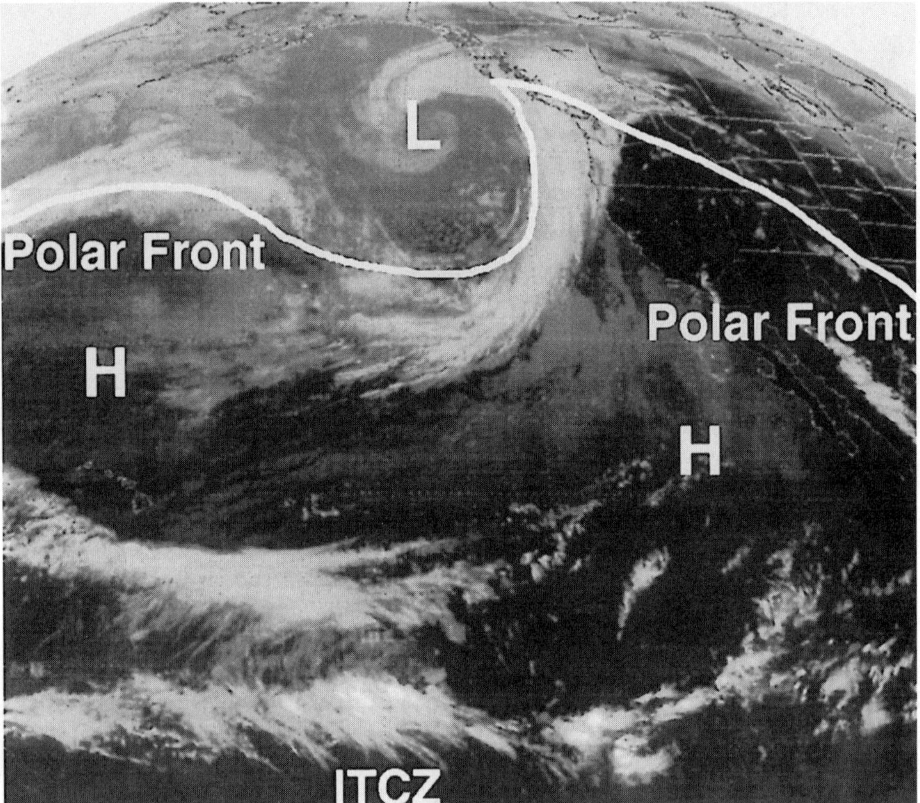

Fig. 3-2. *The components of the general circulation are easily identified on satellite imagery.*

In the northern hemisphere, Coriolis force deflection is to the right; in the southern hemisphere, to the left. Coriolis force is maximum at the poles and zero at the equator. Coriolis force always acts at a right angle to wind direction and directly proportional to windspeed. The effect of Coriolis is to balance pressure gradient force, causing the wind to blow parallel to the isobars. As we shall see, Coriolis force also plays a role in vorticity and the development of hurricanes.

Friction between moving air and the ground slows the wind at and near the surface. This frictional layer is also called the planetary boundary layer. The rougher the terrain, the greater the frictional effect. Over oceans frictional effect is present to about 1000 feet, over land to about 2000 feet. Frictional force always acts opposite to wind direction. As frictional force slows the wind, Coriolis force decreases. However, friction has no effect on pressure gradient force. At and near the surface, the three forces eventually reach equilibrium. Pressure gradient balances the combined effect of frictional and Coriolis forces. Due to the three forces the wind blows across isobars at an angle out of higher pressure toward lower pressure.

As a result, at the surface in the northern hemisphere, wind blows out of high pressure areas toward low pressure area in a clockwise direction and into low pressure areas in a counterclockwise direction. Directions are reversed in the southern hemisphere. This general flow can be seen in Fig. 3-1.

We briefly touched on the transport of air at upper levels with the discussion of Hadley cells. To complete the circulation pattern of major wind systems requires an exchange of air at upper levels. This exchange is required to compensate for, or balance, the movement of air at the bottom of the troposphere.

Strong winds blow in the upper troposphere and lower stratosphere. When these winds become strong enough, they are called a jet stream, as discussed in Chapter 1. Refer to Fig. 3-3. These winds, which generally occur between 30° and 60° latitude, flow more nearly along parallels of latitude than those at the surface.

This wind pattern apparently results from the tendency of winds in large-scale motion systems to retain a constant vorticity about the earth's rotation. The resulting flow, known as *planetary* or *Rossby waves* (planetary scale), have lengths of 2000 to 4000 miles. There are normally three to seven circling the globe at any one time. Their length, amplitude, and position are influenced by differential heating at the surface and mountain barriers, such as the Rockies. One theory holds that the Ice Age was, at least in part, caused by the deflection of the jet stream due to the creation of the Himalayan Mountain range.

Only pressure gradient and Coriolis forces act on upper-level winds. Without frictional force in play, these winds blow parallel to contours as depicted on upper-level

UPPER LEVEL WAVES

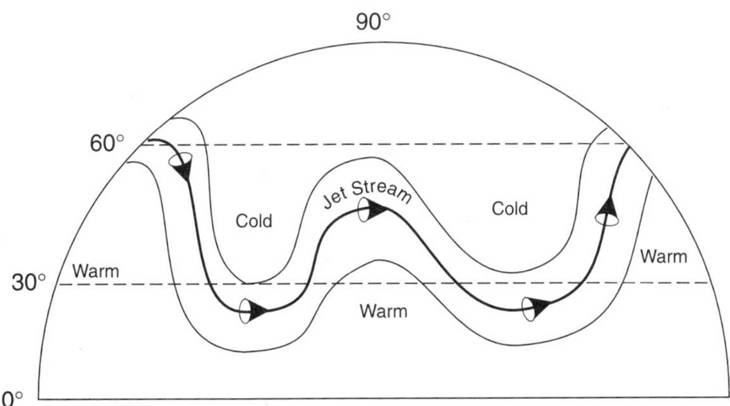

Fig. 3-3. *Strong winds blow in the upper troposphere and lower stratosphere; when these winds become strong enough they are called a jet stream.*

constant-pressure charts. The dashed lines in Fig. 3-3 represent lines of constant pressure—contours—in the upper atmosphere. Upper-level wind flow can be found on constant-pressure charts.

Because the earth's axis is tilted about 23°, the sun's maximum heat strikes different latitudes during the year, causing the seasons. In the northern hemisphere, the general circulation, described in Fig. 3-1, moves south during the winter and north during the summer.

In the summer, the eastern Pacific high blocks weather systems approaching the northwest Pacific coast, forcing them on a more northerly track into Canada. The Bermuda high brings a moist, warm southerly flow to the southeastern United States.

In the winter, the highs migrate southward, allowing storm systems to move from the Pacific into the Pacific Northwest, sometimes reaching into Southern California. Storms continue eastward through the central and southern plains and along the eastern seaboard. As the polar high moves south, it reaches well into the United States several times a year. Occasionally, the polar air can reach as far south as Texas and Florida, bringing subzero temperatures.

The jet stream also moves south and at a lower altitude than in summer. In fact, the jet stream is strongest in spring in the western Pacific and eastern United States because temperature differences are greatest during this period. The location and strength of the jet stream can be found on the 300- and 200-mb constant-pressure charts.

WINDS

Winds, especially near the surface, are a major factor in nonfatal, noninjurious aviation accidents. The most troublesome phase is landing, followed by takeoffs. Crosswinds and gustiness account for the largest percentage of problems. Of these, many occur due to loss of control. However, long landings, downwind or crosswind landings, delayed go-around, and overshoots account for a significant number. Winds aloft, accompanied by poor planning, account for a number of fuel-exhaustion accidents every year. All are preventable.

Mountains, valleys, and water surfaces affect small-scale (subsynoptic) wind systems, which are superimposed on the larger-scale (synoptic) circulations. These include sea and land breezes, mountain and valley winds, downslope winds, and mountain waves. Of these, downslope winds and mountain waves are the most significant to aviation operations. Wind direction and speed is crucial for determining crosswind component. Variable wind direction can make takeoff and landing difficult, even at relatively slow speeds.

Gusts refer to rapid fluctuations in windspeed. To be reported as gusts on METAR, they must vary by 10 knots or more. Gustiness is a measure of turbulence. The greater the difference between sustained speed and gusts, the greater the turbulence. When a sudden increase of at least 15 knots, sustained at 20 knots or more for at least one minute occurs, a squall is reported. Usually associated with thunderstorm activity, *squall* implies severe low-level windshear as well as severe turbulence. Wind direction, speed, and character (gusts or squall) must be considered when determining crosswind component or the advisability of landing at a particular airport.

Winds aloft forecasts provide the pilot with two valuable pieces of information: wind direction and speed, plus temperature. Both significantly affect aircraft operation and performance. Failure to properly consider and apply either can be potentially hazardous.

In spite of their limitations, winds aloft forecasts can never be ignored. Pilots are required by FARs to consider "...fuel requirements..." and are prohibited from beginning a flight either VFR or IFR "...unless (considering wind and forecast weather conditions)..." the aircraft has enough fuel to fly to destination, an alternate if required, still having appropriate fuel reserves. FAR fuel reserve minimums, which do not necessarily equate to "safe," in no way relieve the pilot from keeping careful track of groundspeed and revising the flight plan accordingly.

When reserves are marginal, good operating practice dictates the careful tracking of position and groundspeed. *Marginal* is not necessarily synonymous with legal; in sparsely populated areas, a fuel reserve of 30 minutes with clear weather reported and forecast might be sufficient. But with marginal weather or thunderstorms, and the nearest suitable alternate 35 minutes away, a 30-minute reserve doesn't make any sense. Legal fuel reserves might not be satisfactory with a busted forecast.

The venturi effect of mountains and mountain passes accelerates winds over ridges and through passes, as illustrated in Fig. 3-4. The dashed lines in the figure are *streamlines.* Streamlines show the flow of the winds. On the windward side, the winds are accelerated due to the compressional effect of the mountain range. On the leeward side, windspeed decreases.

Stronger-than-forecast winds should be expected in these areas, especially within 5000 feet of terrain. On a flight from Van Nuys to Fresno, California, the winds aloft

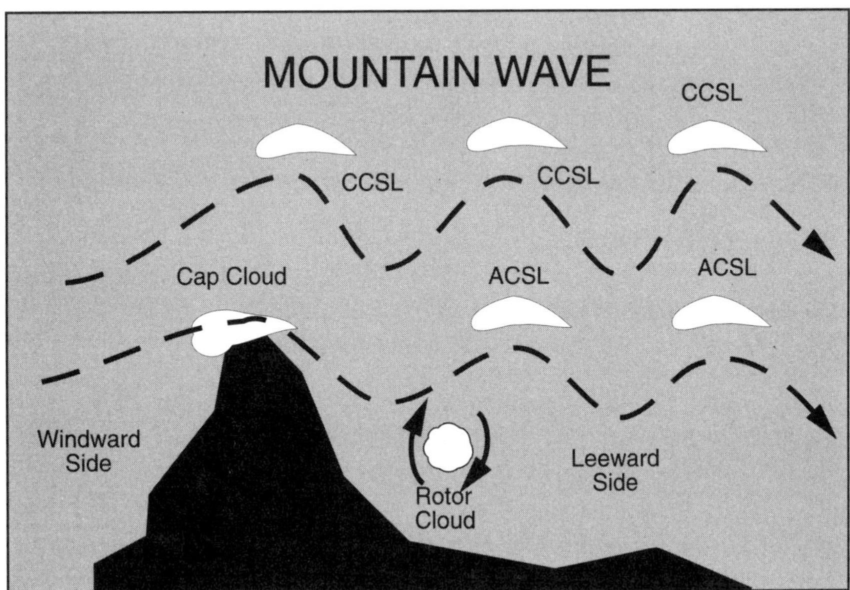

Fig. 3-4. *Major mountain waves occur east of the Cascade, Sierra Nevada, Rocky, and Appalachian mountains.*

were forecast out of the northwest at 20 knots. At 8500 feet in a Piper Arrow with a true airspeed of 130 knots, groundspeed was only 90 knots! The venturi effects of the mountains had doubled the windspeed.

Pilots must continually keep track of groundspeed. Weather updates must be obtained far enough in advance to be acted upon before fuel runs low. Hoping a stronger-than-forecast headwind will abate is folly.

Surface wind conditions can be obtained from METAR reports. Forecast surface winds are available in terminal aerodrome forecasts (TAFs) and the area forecast. Winds aloft forecasts are available in both tabulated or chart form.

Sea and Land Breezes

Recall from Chapter 1 that land surfaces warm and cool more rapidly than water surfaces. Therefore, typically, land is warmer than the sea during the day and cooler than the sea during the night.

During the day the land heats, but the water remains at relatively the same temperature. The land heats the air near the surface through conduction. The air warms, becomes less dense, and rises. The cooler, denser air over the water moves into the relatively lower pressure area. Since the wind blows from the sea to the land, it is called a sea breeze.

With enough moisture and lift, clouds develop at the lifted condensation level over the land. This is particularly true in the southeast United States, with its abundant moisture and unstable air. Activity tends to be widespread, with some areas receiving torrential downpours and adjacent areas remaining dry. Sea breezes, since they occur during the day, can often be seen on visual satellite imagery once clouds develop.

At night this circulation is reversed. The land cools more rapidly than the sea. The wind blows from the cool land toward the warmer water, creating a land breeze. Again, if moisture, lift, and instability are right, thunderstorms and rain showers develop.

Mountain and Valley Winds

In the daytime, mountain slopes become warm, heating the adjacent layer of air. This layer is warmer than air at the same altitude farther from the slope. The resulting density difference creates a convective current in which the air over the valley sinks, forcing the warmer air up the mountains as a valley wind—again so named because the wind blows from the valley toward, or up, the mountain.

At night, the layer of air near the mountain slope cools more rapidly than air over the valley. There is greater air density near the slope than at the same levels some distance horizontally from the slope. The cool air flows down the slope as a mountain wind. The mountain wind—sometimes called a *gravity wind*—often continues to flow down the more gentle slopes of canyons or valleys, and in such cases it takes on the name *drainage wind.* It can become quite strong over some terrain when atmospheric conditions are favorable and, in extreme cases, becomes hazardous when flowing through canyons.

Downslope Winds

Any wind blowing down an incline, where the incline causes the wind, is a *katabatic wind.* Thus a mountain wind is a type of katabatic wind. If the wind is warm, it is called a *foehn* (pronounced "fãn") wind; if cold, it may be a fall wind, such as the *bora.* (A bora is the cold northeast wind on the Dalmatian coast of Croatia and Bosnia/Herzegovina in winter caused by cold air from Russia crossing the mountains and descending over the relatively warm coast to the Adriatic.) A fall wind may also be a gravity wind, such as a mountain wind.

Following its ascent over a mountain barrier, a foehn wind is a katabatic wind. During its movement upslope, it cools at the dry, then moist, adiabatic lapse rate, losing its moisture as precipitation. On descent, it warms through the adiabatic process; the air is warmer than at the same altitude during ascent. Recall our discussion in Chapter 2 and Fig. 2-2.

Strong downslope flows are sometimes given local names, such as the *Chinook* (Native American name for "snow eater") that develops along the eastern slopes of the Rockies. It may raise temperatures by 10°C to 20°C in 15 minutes and can exceed 85 knots. The Chinook often reaches hundreds of miles into the high plains.

The Santa Ana of Southern California is another foehn wind, named for strong winds that blow through Santa Ana Canyon. High pressure over the Great Basin and lower pressure off the Southern California coast cause the Santa Ana. Pressure gradients determine their severity. The stronger the pressure gradient force, the stronger the wind. As in the Columbia River Gorge in Oregon, the winds are accelerated through the passes and canyons. The result is strong gusty winds, severe turbulence, windshear, and crystal-clear skies, except in local areas of blowing dust and sand. The Santa Ana can last from a day or two to a week at a time. These winds continue as long as the pressure gradient is there to support them.

Foehn wall describes the steep leeward boundary of flat, cumuliform clouds formed on the peaks and upper windward sides of mountains during foehn conditions such as the Santa Ana and Chinook winds. This cloud formation should alert pilots to strong winds, possible turbulence, and windshear. The cap cloud in Fig. 3-4 is similar to the foehn wall cloud. When sufficient moisture is present, the cloud forms on the windward side and evaporates on the leeward side of the range.

Mountain Waves

When winds aloft blow in excess of about 40 knots, approximately perpendicular to a mountain range, and speed increases with height in a stable atmosphere, a mountain wave or standing wave can develop. Refer to Fig. 3-4. Turbulence can become severe to extreme. Updrafts and downdrafts occasionally reach 3000 feet per minute and can exist up to the tropopause or slightly higher. Downdrafts may dip to the surface on the leeward side of the mountains. Large waves may extend to the lee of mountains hundreds of miles downstream. Complete overturning, wherein air beneath the wave crest may rotate to just above the surface, producing a rotor cloud with sufficient moisture, may occur under the wave crests at lower levels.

Major mountain waves occur east of the Cascade mountains in Washington and Oregon, the Sierra Nevada mountains in California and Nevada, the Rocky mountains, and the Appalachian mountains. Smaller waves can develop over any hill or mountain.

These waves are termed *standing* because the crests and troughs remain stationary while the wind undulates rapidly through them. With sufficient moisture, clouds form. Although the clouds appear to be stationary, condensation occurs at the leading edge of the cloud, with evaporation at the trailing edge. The clouds are continuously forming and dissipating.

Cap clouds hug the tops of the mountains and appear to flow down the lee side. They are indicators of strong downdrafts on the lee side of a mountain range.

Specific names are given to mountain wave clouds. Standing lenticular stratocumulus (SCSL) is a low cloud; standing lenticular altocumulus (ACSL) is a middle cloud; and standing lenticular cirrocumulus (CCSL) is a high cloud; all indicate mountain wave activity. Lenticular clouds appear smooth and remain stationary to the observer; they might develop as horizontal bands produced by long ridges, as in Fig. 3-5, or circular and stacked from isolated peaks, as in Fig. 3-6. A cap cloud may develop at the mountain crest. Although these clouds imply turbulence, turbulence is not always found.

Lenticular clouds represent visual identification of the wave crests. They may extend to 40,000 feet. Depending on the vertical distribution of moisture in the atmosphere, these clouds may be stacked, as in Fig. 3-6. The distance from the ridge to the first wave crest depends on windspeed, lapse rate, and the shape of the ridge.

Fig. 3-5. *Lenticular clouds appear smooth and remain stationary to the observer.*

Fig. 3-6. *Lenticular clouds develop as horizontal bands produced by long ridges or circular and stacked from isolated peaks.*

Rotor or bell-shaped clouds that often appear as tubular lines of cumulus or fractocumulus clouds parallel to the ridge line underneath the lenticulars always imply severe or greater turbulence. Depending on the strength of the flow and atmospheric conditions, less-developed lines may form downstream beneath the crests of subsequent waves. The base of the rotors is about the same height as the ridge and may extend vertically 3000 to 5000 feet. Violent updrafts are common in the vicinity of rotor clouds.

One such situation occurred at Reno, Nevada, where surface winds were reported gusting to 73 knots. The pilot of a corporate jet reportedly abandoned the approach when all the bottles in the cabin's liquor cabinet broke. (A new definition for severe turbulence?)

Moderate to severe turbulence may be encountered on the windward (updraft) and leeward (downdraft) side of lenticulars. However, the most dangerous situation occurs when the smooth laminar* flow through the wave breaks down, which can result in severe turbulence throughout the vertical depth of the wave. When this occurs, the highest lenticular clouds and, to a lesser extent, the lower wave clouds will have a jagged, irregular edge.

---

*__laminar:__ A nonturbulent flow produced by air moving parallel in adjacent vertical layers.

We encountered a mountain wave in California's Owens Valley, east of the Sierra Nevada, while flying a Cessna 150. With cruise power and attitude, the airplane rode the wave, at the rate of 500 feet per minute, from 8500 feet to 13,500 feet and back down again. The ride was absolutely smooth! Once below 8500 feet, the turbulence was again moderate.

However, mountain waves should never be taken lightly. A Navy T-39 trainer was flying a low-level, high-speed navigational training route in mountainous terrain when it encountered severe turbulence. Gust acceleration loads were so high that aircraft design limits were exceeded, resulting in separation of the tail. All aboard were killed. In addition to the T-39 crash, mountain waves were identified in the crash of a C-118 and in extensive damage to a B-52. While this type of turbulence is obviously critical to traditional low fliers like helicopters, all aircraft are susceptible.

Figure 3-7, a visual satellite image, clearly shows the presence of wave clouds in northern California and Nevada. It illustrates the extent of wave activity. Chances are also good that wave conditions exist in the clear air over the northern Sierra Nevada Mountains and central Nevada.

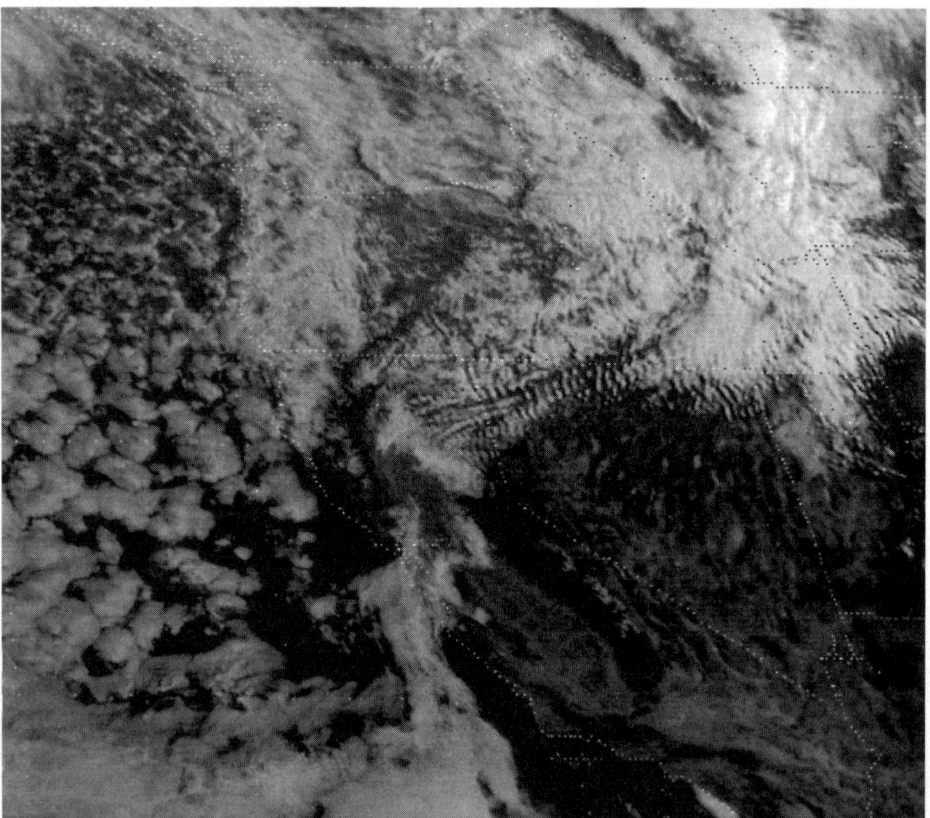

Fig. 3-7. *Visual satellite imagery often clearly shows the existence and extent of mountain wave activity.*

TURBULENCE

Aviation accidents typically are not directly related to turbulence. However, turbulence is a contributing factor to spatial disorientation, loss of control, and structural damage. Therefore, a sound understanding of turbulence, where it is likely to occur, and strategies to reduce its effects are essential.

The intensity of turbulence is, to some degree, affected by aircraft type and flight configuration. U.S. Air Force studies have shown the following to generally increase the effects of turbulence:

- Decreased weight
- Decreased air density
- Decreased wing sweep angle
- Increased wing area
- Increased airspeed

Classifications for the intensity of turbulence can be found in the *Aeronautical Information Manual* (AIM) and *Aviation Weather Services*; however, I prefer the following definitions.

light A turbulent condition during which your coffee is sloshed around but doesn't spill, unless the cup's too full. Unsecured objects remain at rest; passengers in the back seat are rocked to sleep.

moderate A turbulent condition during which even half-filled cups of coffee spill. Unsecured objects move about; passengers in the back seat are awakened by a definite strain against their seat belts.

severe A turbulent condition during which the coffee cup you left on the instrument panel whizzes by the passengers in the back seat. The aircraft might be momentarily out of control, but you don't let on. Anyone not using a seat belt must be peeled off the cabin ceiling.

extreme Usually associated with rotor clouds in a strong mountain wave or a severe thunderstorm, extreme turbulence is a rarely encountered condition in which the aircraft might be impossible to control. The turbulence can cause structural damage. Your passengers are becoming concerned by the beads of sweat on your brow, your white knuckles, and new frequency and new transponder code you have just selected—121.5 and 7700.

Windshear (WS), especially low-level windshear (LLWS), has a significant effect on aviation operations. Windshear is any change in windspeed or direction, either vertically or horizontally, over a relatively short distance. LLWS is defined as windshear that occurs within 2000 feet of the surface. LLWS is divided into two categories: convective and nonconvective. Convective, or thunderstorm-related, LLWS is addressed in Chapter 5, along with our discussion of thunderstorms. Nonconvective LLWS is contained in this chapter.

Pilots should realize there is a difference between windshear and turbulence. Windshear causes airspeed changes in one direction (plus or minus, but not both) with a sustained change in vertical speed. Turbulence, on the other hand, causes airspeed fluctuations (both plus and minus) with no appreciable, or only a momentary, change in vertical speed.

Nonconvective LLWS can be caused by fronts, low-level jet streams, terrain, valley winds, sea breezes, lee side effect, inversions, or Santa Ana or similar foehn-like winds. The occurrence, exact location, and intensity of LLWS are difficult to predict.

Pireps (pilot reports) from air carriers, the military, and corporate aircraft tend to be more accurate because of the pilot's training and experience. Few student pilots fly DC-10s or F-14s. New and low-time pilots (inexperienced) tend to overestimate intensities of turbulence and icing; then they think they've experienced severe conditions and might not heed reports or forecasts. This is not to say pireps from pilots of Cessna 150s or Piper Tomahawks are never accurate. Pireps, along with other reports and forecasts, should never be ignored, but they should be evaluated with all available information. Pilots must remember that, due to its transitory nature, the exact location and intensity of turbulence, like icing, is difficult to forecast.

Mechanical Turbulence

An object placed in any moving air current impedes the flow, causing abrupt changes in wind direction. As the current closes in behind the object, eddy currents develop leeward of the obstruction. This turbulence is caused by the obstruction and not by any meteorological phenomena inherent in the air itself. Turbulence caused in this manner is termed *mechanical.*

Air flowing through mountainous terrain is forced upward on the windward side and spills downward over the leeward side, as illustrated in Fig. 3-4. The degree of turbulence induced by the mountains depends on the shape and size of the mountains, the direction and speed of the wind, and the stability of the air. Downdrafts on the leeward side may be dangerous and can place an aircraft in an attitude from which the pilot may not be able to recover.

As stated in Chapter 1, a recommended strategy is to approach the mountain range at a 45° angle. If the aircraft cannot safely clear the crest, only a 45° turn is required to reach lower terrain.

Surface winds in excess of 20 knots indicate moderate or greater mechanical turbulence, especially over rough terrain. Favorable conditions for turbulence exist just before, during, and after storm passage, especially when winds blow perpendicular to mountain ridges. For example, consider a wind from 230° at 45 knots gusting to 90 knots, which occurred at Mammoth Lakes, California, just after storm system passage, with winds perpendicular to the rugged California Sierra Nevada Mountains.

Runway numbers correspond to their magnetic orientation to the nearest 10°. In the following example the Mammoth runway alignment is east-west (09-27) in relation to magnetic north. Magnetic variation in this area is 15° east. Runway headings are magnetic and METAR winds true; to convert wind direction from true to magnetic, subtract easterly variation (east is least). Therefore, the Mammoth wind is blowing from 215° magnetic (230 − 015 = 215). At an angle of 55° to the runway (270 − 215 = 55), this results in a 35-knot crosswind component for the sustained speed, and 70 knots for the gusts!

Obviously, this airport would not be suitable for landing. For one thing, the highway patrol closed the roads and no one could pick you up after the, umm, arrival. Most aircraft manuals specify maximum demonstrated crosswind component. Every year pilots attempt to test these values. Some pilots even succeed. Each pilot should know his or her limitations and that of the aircraft. As a flight instructor, I always give my students specific crosswind limitations, always with an alternate should they be exceeded.

Uneven surface heating affects the landing approach. Rocky terrain, plowed fields, and paved areas produce predominantly upward currents. These currents force the aircraft above the normal glidepath, resulting in overshooting the touchdown point. Trees, rivers, lakes, and green fields produce predominantly downward currents. These currents allow the aircraft to fall below the normal glidepath, causing it to undershoot the touchdown point. Often both types of currents are present, requiring the pilot to make several corrections on the final approach course.

Thermal Turbulence

Daytime heating causes rising air currents that produce *thermal turbulence,* also called *convective turbulence.* Thermal turbulence usually occurs within 7000 feet of the surface in stable or conditionally unstable air. Vertical movement requires an initiating force, in this case surface heating. In stable or extremely dry air, skies remain clear. Should air parcels reach the lifted condensation level, saturation occurs, and stratocumulus, or fair-weather cumulus, clouds form. These clouds are most often scattered and rarely become overcast, as illustrated in Fig. 3-8. There are rising air currents in the clouds, and descending currents in the clear air. Flight will be turbulent below the clouds and smooth above the clouds. The air is stable since there is little vertical development.

Should the air be conditionally unstable—a parcel of air that becomes unstable on the condition it is lifted to the level of free convection (LFC)—cumuliform clouds form, which can develop into air mass thunderstorms. Figure 3-9 illustrates this situation. Rather than the scattered stratocumulus or fair-weather cumulus of Fig. 3-8, towering cumulus have developed. In this case, expect rainshowers and thunderstorms to form.

Although thermal turbulence rarely becomes severe, it can be extremely uncomfortable and annoying. Because thermal turbulence is caused by surface heating, it can usually be avoided by flying before mid-morning or waiting until late afternoon. Otherwise, the only remedy is to climb above the turbulent layer, which might be marked by clouds. A word of caution to the VFR pilot: If you elect to fly above the clouds, be careful not to get caught on top. Should the air mass be conditionally unstable, clouds can build at an alarming rate and close up even faster.

Flying a new Cessna 150 from Kansas to California, I found myself flying between Winslow, Arizona, and Needles, California, during the afternoon. Skies were clear, and winds aloft were light and variable. I had to climb to 12,500 feet to reach smooth, cool air. Descending into Needles, the turbulence was continuous light to moderate below about 11,000 feet; surface winds were calm.

On another occasion, I was flying one afternoon between Oklahoma City and Amarillo, Texas, in air that was conditionally unstable. As is my habit, I prefer to fly above the

Fig. 3-8. *Rising air currents in the clouds and descending currents in the clear result in turbulence below and smooth air above the clouds.*

clouds in clear, smooth, cool air. The cumulus appeared to top out at about 9000 feet, so, I thought I'd climb to a cruising altitude of 10,500. Something was strange. I was in what appeared to be level flight, but only indicating 60 knots in my Cessna 150. A scan of the instrument panel revealed the problem. I was not in level flight, but in a climb attitude above a sloping cloud deck! Topping the clouds was impossible, and I was forced to descend and bounce the rest of the way to Amarillo at 4500 feet.

Windshear-Induced Turbulence

Windshear-induced turbulence is caused by frontal systems and is associated with clear air turbulence and thunderstorms. Because of their significance, we discuss the latter in subsequent sections. In this section we review three types of windshear-induced turbulence: inversion-induced, evaporative cooling, and Kelvin-Helmholtz (K-H) windshear turbulence.

Inversion-induced windshear turbulence develops along the boundary between cool air trapped near the surface and warm air aloft. The turbulence tends to be strongest in valleys during morning hours when temperature differences are greatest. Moderate or greater turbulence may be encountered penetrating the layer. After an initial outside air temperature rise during the climb through the haze, the air on top will be clear and cool. It is possible to have several haze or smoke layers trapped in inversions aloft. Recall our discussion of haze in the last chapter.

In the Los Angeles Basin, a haze boundary often develops with a Santa Ana condition. Warm, dry desert air overruns haze trapped in cooler, moist marine air. Moderate or greater windshear turbulence can be expected penetrating the transition zone. Takeoff and landing can be hazardous with the boundary close to the runway, often marked by a distinct transition between clear and smoggy air.

A similar condition occurs in the San Francisco Bay area. Strong offshore winds override cool marine air, causing an area of strong windshear. At times the shear line reaches the surface. At the Oakland airport, surface winds for runway 27 can be northeast at 15 gusting to 20 knots and for runway 29 northwest at 5 knots. Needless to say, severe LLWS can develop.

Another type of windshear, which occurs in a vertical plane rather than horizontal like inversion-induced, is evaporative cooling turbulence. Evaporative cooling turbulence develops in areas of precipitation, usually in a dry environment with convective activity. Precipitation evaporates and cools the air, causing downdrafts. A pilot penetrating these areas will encounter windshear turbulence, which can be severe. Turbulence can be avoided by circumnavigating areas of precipitation.

H-K turbulence is caused by windshear. It develops when there is a proper balance of windshear and stability. Stability is important; if the atmosphere is too unstable, mixing occurs and the waves do not develop. With large windspeed changes over a short distance in a stable atmosphere, ingredients are right for H-K turbulence.

Fig. 3-9. *Should the air be conditionally unstable, cumuliform clouds form, which can develop into air mass thunderstorms.*

Under proper conditions an undulating wave forms. If the shear is strong enough, the crests overrun the troughs. With sufficient moisture, a herringbone-shaped cloud develops. Wavelengths are much shorter than in a mountain wave, usually less than 2 miles. The implications are severe windshear and, like the mountain wave, strong updrafts on the windward side of the wave and strong downdrafts on the leeside.

There is evidence to show that H-K windshear and turbulence have either caused or been a contributing factor in several aircraft accidents. Like wave rotor clouds, anytime a pilot sees a cloud indicating an overturning of the air, such as K-H clouds, use caution. This is especially true in mountainous regions. K-H clouds are not a type reported by weather observers; perhaps they should be.

Frontal Turbulence

Frontal turbulence is caused by surface temperature differences exceeding 5 degrees Celsius within 50 miles of the front; it usually occurs below 15,000 feet msl. Since temperature is the determining factor, speed or type of front is not involved in the extent of frontal turbulence. Other types of turbulence, however, such as mechanical or windshear, may also accompany a front. In fact, rapid changes in wind direction and speed below 3000 feet agl within 200 miles of an advancing front may produce low-level windshear.

Clear Air Turbulence

Turbulence encountered in clear air not associated with cumuliform clouds, usually above 15,000 feet and associated with windshear, is classified as clear air turbulence (CAT). Slight, rapid, and somewhat rhythmic bumpiness without appreciable changes in altitude or attitude defines *chop.* Since chop does not cause appreciable changes in altitude or attitude, it would not be severe.

CAT is very patchy and transitory in nature. The dimensions of these turbulent patches are quite variable, generally on the order of 2000 feet in depth, 20 miles in width, and 50 or more miles in length. The patches elongate in the direction of the wind. The dimensions of these areas are on the microscale, and the exact position of specific areas is difficult, if not impossible, to forecast.

CAT is greatest in areas of strongest shear. Moderate or greater CAT occurs most frequently during January and February. Jet stream winds are at lower altitudes, with more frequency and strength during the winter than in summer.

Windspeeds and curvature of contours provide a clue to clear air turbulence. Often the curvature of the contours has more effect on the severity of the turbulence than does windspeed. Areas of potential turbulence occur in

- Sharp troughs
- The neck of cutoff lows
- A divergent flow

Turbulence in these areas can exist despite relatively low windspeeds.

Another factor associated with the jet is windshear turbulence. With an average depth of 3000 to 7000 feet, a change in altitude of a few thousand feet often takes the aircraft out of the worst turbulence and strongest winds. Maximum jet stream turbulence tends to occur above the jet core and just below the core on the north side. Additional areas of probable turbulence occur where the polar and subtropical jets merge or diverge.

At mid latitudes, such as in the United States, the jet stream and areas conducive to CAT can usually be found on the 300-mb and 200-mb constant pressure charts.

Wake Turbulence

Although wake turbulence from larger aircraft is a factor en route, it is a more serious concern during takeoffs and landing. During the takeoff and landing phase, the aircraft is close to the ground and at relatively slow speeds.

A wing producing lift generates a disturbance or wake caused by a pair of counterrotating vortices trailing the wing tips. Vortices occur because of the pressure differential above and below the wing at the tips. After completion of the roll up, the wake consists of two counterrotating cylindrical vortices. The strength of a vortex depends on the weight, speed, and shape of the wing. The greatest vortex strength occurs when the generating aircraft is heavy, slow, and clean—landing gear retracted.

Refer to Fig. 3-10. Vortices develop as soon as the aircraft leaves the ground and cease upon touchdown. The vortex circulation is outward, upward, and around the wing tip when viewed from either ahead or behind. Vortices remain spaced a bit less than a wingspan apart, drifting with the wind, at altitudes greater than a wingspan above the ground. Vortices from larger aircraft sink at a rate of several hundred feet per minute, slowing their descent and diminishing in strength with time and distance behind the generation aircraft. Atmospheric turbulence hastens breakup. When the vortices of larger aircraft sink close to the ground (100 to 200 feet), they tend to move laterally over the ground at 2 or 3 knots. A light wind of 1 to 5 knots could result in the upwind vortex remaining over the runway.

IFR operations are not immune to the hazards of wake turbulence. In one fatal accident, ATC placed a Piper Navajo too close behind a Boeing 727. The Piper was less than 3 miles behind the 727 and above its glidepath when cleared for the approach. Apparently, the Piper was in a steep rate of descent, trying to intercept the glideslope, when it encountered the 727's wake and crashed.

A Twin Commander pilot following a 727 reported severe turbulence. Although the Commander was nearly 5 miles behind the jet, the NTSB believes the pilot most likely lost control due to slow speed and possible icing and wake turbulence.

Pilots can never allow ATC to put them in an untenable position. The Piper pilot should have initiated a missed approach rather than attempt to dive to the glideslope. Situation awareness was also a factor. The proximity to the jet should have been another clue that a missed approach was in order. Apparently the Commander pilot allowed the airspeed to decay well below normal approach speed. This fact, coupled with possible ice and wake turbulence, led to disaster.

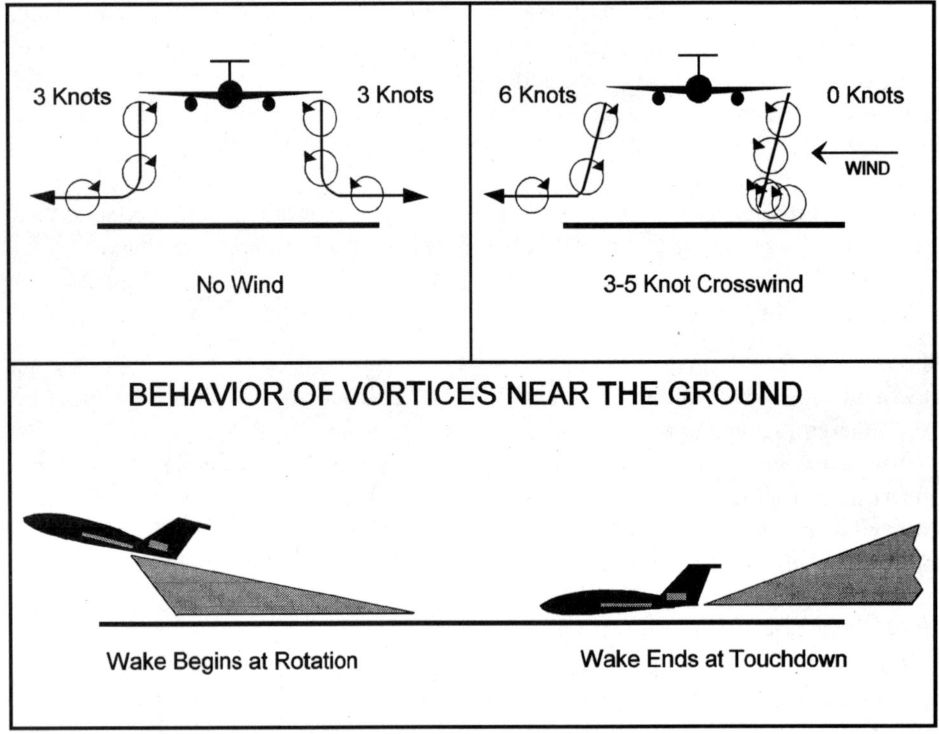

Fig. 3-10. *Vortices develop as soon as the aircraft leaves the ground and cease upon touchdown.*

STRATEGIES

Knowledge is the key to avoiding strong or gusty surface wind accidents. Three areas of knowledge are required. First, know your own limitations. Second, know the performance and limitations of your aircraft. Third, know the surface wind conditions.

What are your personal minimums? In our Air Force Aero Club, student pilots were limited to 10 knots of surface wind, with no more than a 5-knot crosswind component. As an instructor, I impose specific limits based on the individual student's training and experience. For solo cross-country flights, students always had an alternate in case adverse winds developed. When instructing in the Mojave Desert we necessarily had to train in higher winds and crosswinds because of the typically strong winds in that area. It makes no sense to attempt a takeoff or landing with a 25-knot crosswind when you've trained to a maximum 15-knot crosswind.

What is the capability of your aircraft? I have a friend with thousands of hours of military experience. He owns a Kitfox. His personal limits are 15 knots of wind and no more than 10 knots of crosswind component. These limits are based more on the aircraft than his experience.

How can we determine wind conditions? More and more airports have either towers or automated weather reporting systems. If not, check with a nearby field, or observe smoke or trees on the ground. Note wind drift in the pattern and plan accordingly. Finally, if anything isn't right, go around! On approach to Ontario, California, during a Santa Ana wind, a DC-8 made a go-around at about 100 feet when caught by a strong gust. During the takeoff roll, abort if everything isn't going right. The slower the aircraft, the easier it is to control.

Continually updating the weather picture is the key to managing a flight. Winds aloft can be a welcome friend eastbound or a terrible foe westbound. With limited range, even a small change in winds at altitude can have a disastrous result. At the first sign of unexpected winds, Flight Watch should be consulted, if for no other reason than to provide a pilot report. A significant change in wind direction or speed is often the first sign of a forecast gone sour. A revised flight plan might be required. Flight Watch can provide needed additional information on current weather, pireps, and updated forecasts upon which to base an intelligent decision.

The following situation illustrates how a series of small and, at the time, seemingly insignificant factors have the potential to lead to disaster. The flight from Van Nuys, California, to Tonopah, Nevada, was based on four hours of fuel and a 10-knot head wind, time en route estimated 3:15. My Cessna 150 was fueled Friday when I arrived at Van Nuys. During the preflight Sunday morning, I noticed the fuel was not at the top of the filler neck—factor one. This was not unusual because the airplane was parked on a slight incline and some fuel tends to vent overboard. The departure required an IFR climb to on-top conditions, which added about 15 minutes to time en route—factor two.

Over Trona, California, about halfway, groundspeed checks indicated winds were as forecast. Calculations indicated adequate fuel for Tonopah based on four hours of fuel and ignoring the extra time required for departure.

The Cessna 150 climbs like a wet mop, so I decided not to land at Trona—factor three. The fuel gauges were bouncing on zero and I still had 30 minutes to destination and there were no suitable alternates—factor four. I made a straight-in approach and had everything stowed, ready to crash, but I landed safely in spite of some extremely poor planning. By the way, after I landed they put 22.6 gallons in my 22.5-gallon-usable airplane. Never again.

A Grumman Tiger pilot-instructor with a student was not so fortunate. On a flight from Salt Lake City to Tonopah, the plane crashed short of the airport, out of fuel. The pilot couldn't understand why, after calculating the airplane had 2:45 fuel, the engine quit after only 2:31. Needless to say, the FAA wanted to have a little chat with this pilot.

I have on many occasions flown from Southern California to destinations in the Midwest and East. Many of these flights were completed in Cessna 150s. The legs through the intermountain region often stretch the 150's range.

On a leg from Phoenix to Albuquerque, I selected a point, a little over halfway, to make the decision to divert. On this occasion, the promised 10-knot tailwind was as advertised and the flight was completed as planned.

On another flight from Prescott to Albuquerque, things were just not meant to be. Crossing Winslow, Arizona, the Cessna's groundspeed never reached three digits. I changed the flight plan and preceded to Gallup, New Mexico. Hoping that a stronger than forecast headwind will abate is folly. The theme of the preceding examples is to have a plan, and then follow it!

Pilots can avoid the worst effects of turbulence by planning flights above the altitudes of low-level turbulence or avoiding the time of day when turbulence is strongest. To avoid the worst effects of mountain-wave turbulence, pilots should remain at least 5000 feet above the mountain crest. In the west, most light aircraft simply don't have that performance, leaving three options: Select a course with lower terrain, wait it out, or take a chance on getting your fillings knocked loose and maybe losing the airplane. Passenger comfort and safety should be the priority consideration.

If reports or forecasts indicate turbulence, a pilot can minimize the hazard when turbulence is encountered. Turbulence imposes gust loads that appear to be almost instantaneous. Gust loads increase with the speed of the aircraft and gust velocity. The pilot's first task should be to secure loose objects.

If light or moderate turbulence is encountered or expected, avoid flight in the caution range—the yellow arc on the airspeed indicator. If severe turbulence is encountered or expected, reduce to maneuvering speed. Maneuvering speed is not printed on the airspeed indicator, although it's usually placarded in the vicinity of the instrument. Because gust load factor decreases as wing loading increases, maneuvering speed increases with the aircraft's gross weight. Therefore, it's often necessary to determine maneuvering speed based on gross weight.

In moderate or greater turbulence, fly attitude rather than altitude. Disengage the autopilot altitude hold, if in use. The aircraft should already be slowed to turbulent air penetration speed. Don't chase airspeed or altitude. It's like riding a horse—go with it rather than fighting it. The object is to avoid imposing additional abrupt maneuvering loads. For the most part, ignore altitude unless terrain clearance becomes a problem; if VFR, try to avoid IFR cardinal altitudes (4000, 5000, 6000, etc.) or opposite direction VFR altitudes; if IFR, inform air traffic control (ATC) of the problem and any altitude deviations required. ATC increases vertical separation during severe conditions.

We must evaluate pilot weather reports (pireps) within the context of METAR reports, forecasts, and other pireps. A single report of severe turbulence from a Beech Sundowner under clear skies and light winds should be viewed with skepticism. On the other hand, a report of severe turbulence from a Cessna 172 with conditions favorable for a mountain wave and advisories in effect should be taken very seriously. Pireps that are not objective are worse than useless. Not only do they give a false impression to other pilots, but forecasters must take them as fact and issue advisories, which undermine forecast credibility.

A primary reason for high-altitude flying is to avoid mechanical, frontal, and mountain-wave turbulence; however, the flight levels have their own problems—windshear or clear air turbulence. When problems are encountered, Flight Watch can help you find a smooth altitude or alternate route. If the pilot elects to change altitude, an update of actual or forecast winds aloft is often necessary.

Although there are no hard and fast rules as to exactly where turbulence will develop or its intensity, the following table provides an overview of turbulence locations, probability, and intensity.

Table 3-1. Locations and conditions for probable turbulence

Light Turbulence

1. In hilly and mountainous areas even with light winds
2. In and near small cumulus clouds
3. In clear air convective currents over heated surfaces
4. With weak windshear in the vicinity of:
   a. Troughs aloft
   b. Low pressure areas aloft
   c. Jet streams
   d. The tropopause
5. Within 5000 feet of terrain:
   a. When winds are near 15 knots
   b. Where the air is colder than the underlying surface

Moderate Turbulence

1. In mountainous areas with a wind of 25 to 50 knots perpendicular to the ridge:
   a. From the surface to 5000 feet above the tropopause—
      (1) Within 5000 feet of the ridge
      (2) At the base of relatively stable air below the tropopause
      (3) Within the tropopause
2. In and near dissipating thunderstorms
3. In and near towering cumulus
4. Within 5000 feet of the surface:
   a. When surface winds exceed 25 knots
   b. In areas of strong surface heating
   c. At the boundary of a strong inversion
5. In fronts aloft
6. Where:
   a. Vertical windshear exceeds 6 knots per 1000 feet
   b. Horizontal windshear exceeds 18 knots per 150 miles

Severe Turbulence

1. In mountainous areas when wind exceeds 50 knots perpendicular to the ridge:
   a. Within 5000 feet of the ridge
   b. At and below the ridge in rotor clouds or rotor action
   c. At the tropopause
   d. At the base of a stable layer below the tropopause
   e. Extending outward on the lee of the ridge for 50 to 150 miles
2. In and near growing and mature thunderstorms

Table 3-1. (Continued)

Severe Turbulence

3. In towering cumulus clouds
4. Fifty to 100 miles on the cold side of the center of the jet stream, in troughs aloft, and in lows aloft when:
   a. Vertical windshear exceeds 6 knots per 1000 feet
   b. Horizontal windshear exceeds 40 knots per 150 miles

Extreme Turbulence

1. In mountain waves, in and below the level of well-developed rotor clouds, sometimes extending to the ground
2. In growing severe thunderstorms, especially squall lines, with:
   a. Hailstones ¾ inch or greater
   b. Strong radar echoes
   c. Continuous lightning

A wake turbulence encounter can result in one or two jolts of varying severity or can be catastrophic. The probability of induced roll increases when the encountering aircraft's heading is generally aligned with the flight path of the generating aircraft. The key to wake turbulence avoidance is knowledge and defensive flying. Avoid the area below and behind the generating aircraft, especially at low altitude where even a momentary encounter could be hazardous. Pilots should be particularly alert in calm wind conditions and situations where the vortices could

- Remain in the touchdown area
- Drift from aircraft operating on a nearby runway
- Sink into the takeoff or landing path from a crossing runway
- Sink into the traffic pattern from other airport operations
- Sink into the flight path of VFR aircraft

The following vortex avoidance procedures are recommended:

- *Departing behind a larger aircraft:* Note the larger aircraft's rotation point and rotate prior to that point. Continue climb above the larger aircraft's climb path until turning clear of its wake. Avoid subsequent headings that cross below or behind a larger aircraft.
- *Intersection takeoffs on the same runway:* Be alert to adjacent larger aircraft operations, particularly upwind of your runway. If an intersection takeoff clearance is received, avoid subsequent headings that cross below a larger aircraft's path.
- *Departing or landing after a larger aircraft executing a low approach, missed approach, or touch-and-go landing:* Because vortices settle and move laterally

near the ground, the vortex hazard may exist along the runway and in your flight path after a larger aircraft has executed one of these maneuvers, particularly with a light quartering wind. You should ensure that an interval of at least two minutes has elapsed before your takeoff or landing.

- *Landing behind a larger aircraft on the same runway:* Stay at or above the larger aircraft's final approach flight path, note its touchdown point, and land beyond that point.
- *Landing behind a larger aircraft when a parallel runway is closer than 2500 feet:* Consider possible drift to your runway. Stay at or above the larger aircraft's final approach flight path and note its touchdown point.
- *Landing behind a larger aircraft on a crossing runway:* Cross above the larger aircraft's flight path.
- *Landing behind a departing larger aircraft on the same runway:* Note the larger aircraft's rotation point and land well before its rotation point.
- *Landing behind a departing larger aircraft on a crossing runway:* Note the larger aircraft's rotation point. If it is past the intersection, continue the approach and land prior to the intersection. If the larger aircraft rotates prior to the intersection, avoid flight below the larger aircraft's flight path. Abandon the approach unless a landing is ensured well before reaching the intersection.
- *En route:* Avoid flight below and behind a large aircraft's flight path. If a larger aircraft is observed above on the same track (meeting or overtaking) adjust your position laterally, preferably upwind.

The flight from Van Nuys to Long Beach was on a typical L.A. Basin day. Above the haze there was clear, smooth air. I spotted a "three-holer" (Boeing 727) on approach to Los Angeles International. It was above my flight path, so I planned to cross about 3 miles behind. I penetrated the wake at a 90° angle. It grabbed the airplane and almost instantaneously dropped me about 50 feet. Because I was belted in, I was also yanked down. My recollection of the incident was seeing my notepad flying out of my pocket in front of my face and two ashtrays crossing in front of me dumping butts all over the place. I remember the first thing I did was check to see if the tail was still there. It was.

Wake turbulence encounters with the Boeing 757 have received a lot of press lately. A Boeing 737 captain said a 757 wake rolled his aircraft 45° on approach to Salt Lake City. That must have been exciting. Keep in mind that wake turbulence from much smaller aircraft, including helicopters, can significantly affect small aircraft.

Because of the hazards of wake turbulence, controllers are required to apply specific minimum separation between aircraft of various weight categories in the form of time or distance. For example, a two-minute interval is provided for a small airplane departing behind a heavy airplane, or 6 miles for a small airplane landing behind a heavy jet. Keep in mind that pilots have the option to request additional wake turbulence separation. Like icing and thunderstorms, avoidance is the key dealing with the wake turbulence hazard.

4
Signpost in the sky

PILOTS CAN DETERMINE QUITE A LOT ABOUT THE STATE OF THE ATmosphere from clouds, specifically from cloud types. We have already discussed some cloud types and their significance on aviation operations—fog, low ceilings, and mountain waves.

Cloud types are divided in four categories: low, middle, high, and those with vertical development. However, for aviation purposes, other than some specific types, such clouds fall into two major divisions: stratiform and cumuliform. We discuss the significance of each. Before leaving this subject, we discuss effects of terrain on clouds and precipitation, for example, lake effect.

Next we discuss aircraft icing. Icing intensities, types, and causes are presented. Structural icing affects aircraft in three ways: flight and engine instruments, induction systems, and structural icing. An erroneous assumption is that icing only affects the IFR pilot. This is not true. Icing affects the VFR pilot as well as the IFR pilot. In fact, as we shall see, based on total numbers VFR pilots are involved in more icing-related accidents than IFR pilots. Therefore, our discussion applies equally to both types of operations. The last item in this section is runway ice. Snow- or ice-covered runways can have a significant effect on taxi, takeoff, and landings. Pilots that fly in these conditions must be prepared for these situations. Again, this section applies to both VFR and IFR operations.

The chapter concludes with strategies to avoid or counter the effects of icing. For most light aircraft, avoidance is the only solution. Once the engine has failed due to carburetor ice or the airfoils incapacitated by ice, the pilot has few, if any, options. Through personal and accident scenarios and a sound understanding of the subject, we will learn to preclude or avoid hazardous icing situations.

CLOUD TYPES

Clouds form through a cooling process. The process initiates and then must sustain condensation or sublimation. Cooling processes are adiabatic or diabatic. We have previously discussed the adiabatic process. The *diabatic process* involves the exchange of heat with an external source.

The adiabatic process cools the air by raising it through the processes of convection, convergence, or orographic lifting. The diabatic process produces a loss of heat. The loss may occur through terrestrial radiation, resulting in fog or low clouds. Conduction through contact with a cold surface may result in dew, frost, or fog. The process may be associated with the movement of air across a cold surface—advection. Finally, the process may occur through mixing with colder air. If the mixture has a temperature below its dewpoint, clouds or fog may form.

There are a number of methods of cloud classification. Clouds may be classified according to appearance, how they are formed, or the height of their bases. It was not until 1803 that Luke Howard, an Englishman, first classified cloud forms. He divided clouds into three main categories using Latin names:

- Cirrus, meaning curly
- Stratus, meaning spread out
- Cumulus, meaning heaped up

Two prefixes/suffixes may be added:

- Alto, meaning high
- Nimbo, meaning rain

Today, meteorologists divide clouds into four main groups:

- Low clouds (bases near the surface to about 6500 feet)
- Middle clouds (bases from 6500 feet to 20,000 feet)
- High clouds (bases at or above 20,000 feet)
- Clouds with vertical development (bases near the surface, tops of cirrus)

Sometimes clouds are classified into one of two general classifications: *stratiform* and *cumuliform.* Stratiform describes clouds of extensive horizontal development associated with a stable air mass. Stratiform clouds consist of small water droplets. The following cloud types are classified as stratiform:

- Stratus
- Stratocumulus
- Nimbostratus

- Altostratus
- Cirrostratus

Cumuliform describes clouds that are characterized by vertical development in the form of rising mounds, domes, or towers, associated with an unstable air mass. Because of upward-moving currents, cumuliform clouds can support large water droplets. In the case of cumulonimbus, updrafts can support hail. The following cloud types are classified as cumuliform:

- Altocumulus
- Cirrocumulus
- Cumulus
- Cumulonimbus

Table 4-1 shows the general differences between stratiform and cumuliform clouds. Notice how there is a correlation between Table 4-1 and Table 2-1, Characteristics of Atmospheric Stability.

Table 4-1. Characteristics of Cloud Types

Stratiform	**Cumuliform**
Stable lapse rate	Unstable lapse rate
Poor visibility	Good visibility
Smooth	Turbulent
Widespread cloud mass	More localized cloud mass
Steady precipitation	Showery precipitation
Rime icing	Clear icing

Although not one of the two previously mentioned general classifications, a third generic cloud type may be used. It describes the entire group of high clouds—*cirroform.* Cirroform is often used during pilot weather briefings to translate one or all of the high cloud types: cirrus, cirrostratus, or cirrocumulus.

Figure 4-1 illustrates four cloud types and two cloud groups. Low stratus is in the valleys topping a surface-based inversion, indicating a stable air mass at lower levels. Notice the poor visibility in the valley. Moderate winds over the mountains have produced stratocumulus. The presence of stratocumulus indicates some instability, with good visibility and turbulence, over the mountains. High-level moisture has produced cirrus and cirrostratus—stable air at high levels.

Low Clouds

Stratus clouds indicate a stable air mass. Stratus may lower, becoming fog, after sunset due to the absence of surface heating and decreased winds. During the morning, fog may

Fig. 4-1. *Cloud types reveal stable air in the valleys and at high levels; stratocumulus clouds over the mountains indicate some instability and turbulence.*

lift, due to surface heating, to become a stratus layer. Ceilings and visibilities are typically poor—recall Fig. 4-1. Flight through stratus clouds is smooth. Precipitation, when it occurs, is usually light, often in the form of drizzle.

Stratocumulus clouds represent a moist layer with some convection. Stratocumulus may form from the spreading out of cumulus, which indicates decreasing convection. Stratocumulus can develop from stratus with winds of moderate to strong intensity, as shown over the mountains in Fig. 4-1. Updrafts and turbulence develop below and within the layer.

Fractostratus and *fractocumulus* (also known as *scud,* or shreds of small detached clouds moving rapidly below a solid deck or higher clouds) are normally associated with bad weather. Figure 4-2 shows a layer of scud. Pilots who fly low in poor weather conditions associated with these clouds are known as "scud runners." This practice has given rise to the aviation axiom: "There are old pilots and there are bold pilots, but there are no old, bold pilots." The life expectancy of scud runners is rather poor. An unfortunate pilot attempted to negotiate Oakland's east bay hills under these conditions. He took his two grandchildren with him. As an ex-army aviator friend of mine relates quite frequently: "Cowardice is the better part of valor." By the way, John, my ex-army aviator, is one of the best pilots I have ever flown with.

Nimbostratus are low clouds, usually uniform and dark gray in color. Nimbostratus usually evolve from altostratus that have thickened and lowered, sometimes with a ragged appearance. These clouds are the ordinary rain clouds that produce light to moderate, steady precipitation. Flight in, and in the vicinity of, nimbostratus is usually smooth. The presence of nimbostratus indicates a stable air mass.

Middle Clouds

Middle clouds fall into two general types: altostratus and altocumulus. *Altostratus* clouds indicate a stable atmosphere at mid levels. Some altostratus are thin and semitransparent, while others are thick enough to hide the sun or moon. Figure 4-3 shows a thickening band of altostratus. The sun backlights the cloud in the upper center of the photograph. As in this case, these altostratus clouds indicate the approach of a warm front. Ceilings are high and visibilities are good below the cloud because no precipitation is falling. These clouds can produce precipitation in the form of rain or snow, even heavy snow at times.

Altocumulus clouds indicate vertical motion and instability at mid levels. Altocumulus may be thin, mostly semitransparent. Some altocumulus are thick, developed, and may be associated with other cloud forms. In Chapter 3 we discussed altocumulus standing lenticular clouds associated with mountain waves. Figure 4-4 shows a thick band of altocumulus below a band of cirrus. This cloud often signals the approach of a cold front.

Altocumulus castellanus (ACC), a mid-level cloud, indicates moisture and instability at this level. ACC might indicate thunderstorm development. Showers falling from these clouds can evaporate before reaching the surface, as illustrated in Fig. 4-5. This phenomena is known as *virga*. Virga is rain that evaporates before reaching the surface.

Fig. 4-2. *Scud is normally associated with bad weather.*

Fig. 4-3. *A thickening band of altostratus indicate stable air at mid-levels.*

Fig. 4-4. *Thickening and lowering altocumulus may signal the approach of a cold front.*

Evaporative cooling turbulence develops in the vicinity of virga. Recall from Chapter 3 that precipitation evaporates and cools the air, causing downdrafts and windshear. A pilot penetrating these areas will encounter windshear turbulence, which can be severe. Turbulence can be avoided by circumnavigation of these areas.

Figure 4-6 shows altocumulus castellanus that have developed above a stratus layer. The lower layer of the atmosphere is stable, as indicated by the stratus and haze. The upper layers are unstable, as testified by the ACC. This phenomenon could be hazardous to pilots flying below the stratus and suddenly encountering heavy rain and turbulence. A group of hot-air balloons were downed in California's Napa Valley under similar circumstances.

Altocumulus floccus is a cloud with a cumuliform or rounded appearance. The lower portion is ragged and often accompanied by virga. Altocumulus floccus may evolve from altocumulus castellanus. Like altocumulus castellanus, they indicate moisture and instability at mid-levels in the atmosphere.

High Clouds

High clouds are known as cirrus, cirrostratus, and cirrocumulus. Cirrus clouds are composed entirely of ice crystals. Usually the air is so cold that they present no serious icing hazard.

Fig. 4-5. *Virga is rain that evaporates before reaching the ground.*

Fig. 4-6. *Altocumulus castellanus indicate instability aloft; a stratus layer attests to a stable layer at the surface.*

Cirrus clouds often consist of filaments, commonly known as mares' tails. Other cirrus are associated with cumulonimbus clouds. A thickening cirrus layer may indicate the approach of a front. Figure 4-7 shows a band of cirrus over a stratus deck.

At times, ice crystals or snowflakes can fall from cirrus clouds. As they fall into dry air, they sublimate—change directly from a solid to a gas. These dangling white streamers are called fall streaks.

Cirrostratus describes sheets or layers of cirrus clouds. Sun or moon halos may occur, as illustrated in Fig. 4-8. When cirrostratus appear within a few hours after cirrus in mid latitudes, there is a good probability of an approaching front. Thick bands of cirrostratus often mark the location of the jet stream. Jet stream cirrus can be seen on satellite imagery and are often observed in the area of an upper-level ridge. The poleward edge of the cirrus shield ends in a line parallel to the jet axis.

Cirrocumulus clouds indicate vertical motion at high levels and might indicate high-altitude turbulence. Figure 4-9 shows a bank of cirrocumulus clouds. As discussed in Chapter 3, cirrocumulus standing lenticular clouds often develop in a mountain wave. Transverse lines are often observed in jet stream cirrus. These cloud patterns appear as small-scale lines at an angle almost perpendicular to the jet. These lines look somewhat like waves but are much more irregular than mountain waves. Cirrus streaks are also associated with the jet stream. However, cirrus streaks are parallel to jet core.

Fig. 4-7. *Cirrus clouds often consist of filaments, commonly known as mares' tails.*

Fig. 4-8. *When cirrostratus appear within a few hours after cirrus, there is a good probability of an approaching front.*

Fig. 4-9. *Cirrocumulus indicate vertical motion at high levels and might indicate high-altitude turbulence.*

Cirrus, of itself, has no significance to low-level flights. Think: Extensive shields of cirrus may prevent or lessen the effect of thermal turbulence. As previously mentioned, cirrus is often associated with the jet stream and high-altitude turbulence. Cirrus that form as transverse lines or cloud trails and cirrus streaks indicate moderate or greater turbulence. These clouds might be reported as cirrocumulus. Jet stream cirrus, transverse lines, and cirrus streaks are easily identified from satellite imagery.

A *contrail,* or condensation trail, is a cloudlike streamer frequently observed to form behind aircraft flying in clear, cold, moist air. Contrails typically form in the upper troposphere. Figure 4-10 shows an aircraft contrail over West Virginia. Contrails can form from the addition of water vapor produced by engine combustion. Or, in air that is almost fully saturated, aerodynamic forces around the propeller tips or wings can cool the air and induce condensation.

In a very rare occurrence, an aggregate of ice crystals form in the stratosphere. These clouds are known as mother of pearl.

Clouds with Vertical Development

Clouds with vertical development are *cumulus* and *cumulonimbus.* Some cumulus describe fair weather, seemingly flattened, with little vertical development, no significant weather, turbulent below the bases and smooth on top. Other cumulus contain considerable vertical development, generally towering. This type precedes the development of cumulonimbus clouds and thunderstorms. METAR reports may contain TCU (towering cumulus) to describe this cloud. This designation refers to growing cumulus that resemble a cauliflower, but with tops that have not yet reached the cirrus level. Figure 4-11 shows towering cumulus developing during the summer over California's Sierra Nevada mountains. Note the bubbling, cauliflower appearance of the tops.

With cumulus clouds, expect usually good ceilings and visibilities, except in the vicinity of precipitation. These clouds imply turbulent flying conditions. Precipitation is showery, and heavy at times.

Cumulonimbus clouds exhibit great vertical development with tops composed, at least in part, of ice crystals. Tops no longer contain the well-defined cauliflower shape of towering cumulus. Cumulonimbus may develop a clearly fibrous (cirroform) top, often anvil-shaped. Regardless of vertical development, a cloud is classified as cumulonimbus only when all or part of the top is transformed, or is in the process of transformation, into a cirrus mass. Any cumulonimbus cloud should be considered a thunderstorm with all the ominous implications.

Cumulonimbus mamma—also know as mammatus (CBMAM), from the Latin word meaning udder or breast—result from severe updrafts and downdrafts. They are characterized by lobes that protrude from the bottom of the cloud, as illustrated in Fig. 4-12. They indicate probable severe or greater turbulence and often appear just before or at the beginning of a squall. Avoid these areas.

Fig. 4-10. *Contrails typically form in the upper troposphere.*

Fig. 4-11. *Towering cumulus are growing cumulus that resemble cauliflower but with tops that have not yet reached the cirrus level.*

Fig. 4-12. *Cumulonimbus mamma indicate probable severe or greater turbulence and often appear just before or at the beginning of a squall.*

Heavy, showery precipitation falls from cumulonimbus clouds. (Showers are characterized by the suddenness with which they start and stop and rapid changes of intensity.) Outside the areas of precipitation, the air is usually clear and relatively smooth in the unstable air mass. In the vicinity of the showers, expect severe turbulence and strong up- and downdrafts. Figure 4-13 illustrates these conditions. Notice how visibility is good outside of the rain showers. However, the mountains are partly to completely obscured on the left side of the photograph. Even in the absence of clouds, precipitation is so heavy that VFR flight is not possible, not to mention the turbulence and windshear.

Three additional cloud types associated with convective activity that pilots should be familiar with may appear in the remarks of METAR reports. A *shelf cloud,* layered and resembling shelves, can appear under a thunderstorm. A *wall cloud,* usually on the southwest edge of the thunderstorm, has a lowered base and indicates the storm might be severe. A *roll cloud* appears as a detached, dense, horizontal cloud at the lower front part of the main cloud. All three indicate thunderstorms and the potential for severe weather.

Occasionally a severe thunderstorm breaks through the tropopause into the stratosphere. This is one of the few instances when clouds appear in this layer of the atmosphere.

The contractions "TCU" and "CB" appear in body of METAR reports and "CB" appears on TAFs to indicate the presence of these clouds. Manual METAR observation may contain other significant cloud types, as discussed in this section.

Fig. 4-13. *Heavy showers, not to mention the turbulence and windshear, even in the absence of clouds, may preclude VFR flight.*

Chapter Four

PRECIPITATION

We've used the term *precipitation* throughout the book thus far. I think now we should ensure that we all have the same understanding of this phenomena. Precipitation is any or all of the forms of water particles, whether liquid or solid, that fall from the atmosphere and reach the ground. It is a major classification, but it does not include clouds, fog, dew, or frost. Precipitation is distinguished from cloud, virga, and fall streaks in that it must reach the ground. Precipitation includes

- Drizzle
- Rain
- Freezing drizzle and freezing rain
- Snow grains
- Snow
- Snow pellets
- Ice crystals
- Ice pellets
- Hail

Drizzle is very small, numerous, and uniformly dispersed water drops that may appear to float. Unlike fog droplets, drizzle falls to the ground. Drizzle usually falls from stratus clouds and indicates a relatively shallow cloud layer. It usually takes a cloud thickness of 4000 feet to produce precipitation. Drizzle restricts visibility to a greater degree than rain because it falls in stable air often accompanied by fog, haze, and smoke.

Rain is precipitation in the form of liquid water drops. Rain drops have diameters greater than 0.5 millimeters (mm), which distinguishes them from drizzle drops, which have a diameter less than 0.5 mm. (Right, we're all going to go out and measure the stuff!) From an observational point of view, other than size, drizzle appears to float or fall very slowly, while rain has a definite vertical speed.

We have already discussed how rain reduces forward visibility, but visibility to the side and downward remains good. In moderate rain, forward visibility can be reduced to less than VFR. If the rain is heavy, it can obscure terrain. Flight through rain tends to be turbulent. Avoiding areas of rain not only improves forward visibility, but results in a smoother ride.

Freezing rain and *freezing drizzle* are caused by liquid precipitation falling from warmer air into air that is at or below freezing. Droplets freeze upon impact. Structural icing can be expected while flying through freezing precipitation. Freezing precipitation is probably the most dangerous of all icing conditions. It can build hazardous amounts of ice in a few minutes and is extremely difficult to remove. Freezing rain can flow back along the aircraft, covering the static port, with the resultant loss of accurate pitot-static instruments—airspeed, altimeter, and vertical speed.

Snow grains are small, white, opaque grains of ice, the solid equivalent of drizzle.

Snow is composed of white or translucent ice crystals, chiefly in complex branched hexagonal form and often integrated into snowflakes. Snow can fall about 1000 feet below the freezing level before melting. Snow can often begin with temperatures of 2°C; it's even possible to see snowflakes at temperatures around 10°C. This phenomenon only occurs when the air is very dry. As snow falls into above-freezing air, it begins to melt. The water evaporates and cools the air. Evaporation cools the snow, which retards melting. Water vapor is added to the air, which increases dewpoint. Finally, the air cools and becomes saturated at 0°C.

Dry snow does not lead to the formation of aircraft structural ice. However, *wet snow*—snow that contains a great deal of liquid water—produces structural icing.

Snow flurry is a popular term for snow showers, particularly of a very light and brief nature. On the other hand, a *blizzard* is a term used to describe a severe weather condition characterized by low temperatures, strong winds, and large amounts of snowfall. The term blizzard is thought to have originated in Virginia, but it is now applied to similar occurrences in other countries. In North America, blizzards occur with the northwesterly winds in the rear of low pressure areas in winter. In popular usage, the term often refers to any heavy snowstorm accompanied by strong winds.

Ceiling and visibilities can be good in light snow. Heavy snow, however, can reduce ceiling and visibility to zero and produce whiteout effect.

Snow pellets—small, white, opaque grains of ice—form when ice crystals fall through supercooled droplets and the surface temperature is at or slightly below freezing. Falling from cumuliform clouds, snow pellets are more prone to cause structural icing than snow grains.

Ice crystals might appear to be suspended and fall from a cloud or clear air. They frequently occur in polar regions in stable air and only at very low temperatures. Ice crystals are not assigned an intensity. I was on watch at the Lovelock, Nevada, FSS on a very cold, clear day. Ice crystals were sublimating right out of the clear air. It was a beautiful sight but of absolutely no significance to aviation or anything else.

Ice pellets, formerly *sleet,* are grains of ice consisting of frozen raindrops or largely melted and refrozen snowflakes. They fall as continuous or intermittent precipitation. Ice pellet showers are pellets of snow encased in a thin layer of ice formed from the freezing of droplets intercepted by the pellets or water resulting from the partial melting of the pellets. Ice pellets do not bring about the formation of structural ice, except when mixed with supercooled water. Frequently, ice pellets or ice pellet showers indicate areas of freezing rain above.

Hail is precipitation in the form of balls or irregular lumps of ice, always produced by convective clouds, nearly always cumulonimbus. An individual ball is called a *hailstone.* Thunderstorms that are characterized by strong updrafts, large liquid water content, large cloud-drop size, and great vertical height are favorable to hail formation. The violent updrafts keep hailstones suspended for several up and down cycles. Each cycle adds a layer to the hailstone until it can no longer be suspended in the cloud.

Hail can cause severe damage to objects on the ground as well as aircraft. Blunted leading edges, cracked windscreens, and frayed nerves are a common result of a hail encounter. Like most thunderstorm hazards, avoidance is the only solution.

On a flight from Lancaster's Fox Field to Van Nuys in California, we flew the Cessna 172 into a big, black, ugly cloud. Thunderstorms were not forecast. To this day I'm not sure if it was small hail that we encountered. The echo within the cabin was deafening. It sounded like the windscreen was about to implode, and I was sure it would, at the very least, have to be replaced. We survived and the windscreen was OK, but I don't want to see anything like that again. The solution: Don't fly into big, black, ugly clouds!

We have already discussed how terrain affects the weather as it relates to fog and turbulence. The land and water surfaces underlying the atmosphere significantly affect cloud and precipitation development. As moist air moves upslope and cools adiabatically, condensation produces clouds and precipitation. Showers occur if the air is unstable; when the air is stable, precipitation is more widespread and steady. As the air moves over the crest and downslope, it heats adiabatically; precipitation ceases and clouds dissipate.

One effect of terrain is "rain shadows." This phenomenon predominately occurs in the West. Pacific storms shed most of their moisture over the Cascades of Washington and Oregon and the Sierra Nevadas of California, resulting in semiarid climates east of these ranges. After frontal passage in the western United States, there is typically enough wind and moisture to produce mountain obscurement and trap low clouds where the wind flows up the mountains. This condition may occur for a day or two after frontal passage (Fig. 4-14). Air flowing up from California's Central Valley has caused clouds to develop along the west slopes of the Sierra Nevadas. Because the air is slightly unstable and the wind moderate, stratocumulus have developed. If the air were stable and the winds light, a condition more like the upslope of the western planes would occur.

Fig. 4-14. *Air flowing up from California's Central Valley has caused stratocumulus to develop along the west slopes of the Sierra Nevada.*

Another phenomenon is *lake effect.* Often in winter, cold air moves over relatively warm lakes. The warm water adds both heat and water vapor to the air. The added heat makes the air unstable, resulting in showers to the lee of the lakes. Since it is winter, snow showers develop downwind. These snow showers can be heavy and produce severe aircraft icing. Lake effect snow often occurs in the Great Salt Lake area of Utah and around the Great Lakes. In November 1996, severe lake effect caused heavy snow and the closing of Cleveland's Hopkins International airport for days. During the period several aircraft slid off the runway.

Takeoffs and landings on runways with standing water, slush, or snow can be extra hazardous. Takeoffs should not be attempted in standing water or wet snow greater than ½ inch in depth that covers an appreciable part of the runway. Additional ground roll is required for takeoffs. A soft-field technique should be considered; always follow the aircraft manufacturer's recommendations.

Type and intensity of precipitation are found in the body of METAR reports and forecast in TAFs and the area forecast. METAR remarks may include significant types of precipitation, such as WET SNOW.

ICING

Icing affects airframes, engines, propellers and rotors, and aircraft flight instruments and radios. Supercooled water drops produce airframe icing. These drops are visible, liquid water with temperatures at or below 0°C. When these droplets hit an aircraft, they freeze. Icing can also occur with water droplets greater than 0°, if the airframe temperature is at or below zero, such as on an aircraft descending from an area of cold air. Aerodynamic cooling can also lower the temperature of an airfoil to 0°C even though the ambient temperature is a few degrees warmer. When the temperature reaches −40°C or less, it is generally too cold for airframe ice to form.

Ice on an airfoil disrupts the smooth flow of air; it decreases lift and increases drag. Lift may be decreased by as much as 50% and drag increased by as much as 35%. Ice increases the weight of the aircraft and may affect the engine, reducing power output, or in extreme cases, causing engine failure.

Ice forms on propellers and in jet engine inlets. Even a small amount of ice, if not evenly distributed, can cause stress on the piston engine mounts and propeller. When the propeller sheds ice, a momentary increase in vibration and stress occurs. This can be very exciting as chunks of ice hit the fuselage. Finally, ice affects aircraft intakes and carburetors.

Ice can affect aircraft instruments. (Recall our discussion of the pitot-static system in Chapter 1 and Fig. 1-6.) Iced-over pitot-static instruments can cause false readings or render the instruments useless. Clogging of the pitot tube by ice affects the airspeed indicator only. Many aircraft are equipped with an alternate static source, vented inside the cabin, for emergency use. Static pressure inside the cabin is usually lower than outside static pressure. Therefore, the altimeter reads higher than normal; indicated airspeed is greater than normal; and the vertical speed indicator shows a momentary climb, then operates normally.

Instrument icing has caused jet air carrier accidents as well as general aviation accidents. A Boeing 727 was lost due to an iced-over pitot tube. As static pressure decreased during the climb, the airspeed indicator showed speed increasing. The autopilot attempted to hold airspeed by increasing pitch, resulting in a stall. A Boeing 737 crashed because of an iced-over engine power sensor. The airplane was simply not developing takeoff power, even though the instrument indicated so.

Ice also affects radio communications by reducing antenna efficiency or causing antennas to break off.

Ice, especially on aircraft without ice protection equipment, can obstruct the pilot's view. Needless to say, it's extremely dangerous attempting to land with an iced-over windscreen. Now add the fact that ice can disrupt the function of control surfaces, reduce the effectiveness of brakes, and interfere with landing gear operation.

Therefore, ice adversely affects an aircraft by

- Increased drag
- Loss of lift
- Increased weight
- Reduced power or loss of power
- Interference with control surfaces
- Reduced effectiveness of brakes
- Interference with landing gear operation
- Increased vibration and structural stress
- Reduced or precluded forward vision
- Loss of, or false, instrument readings
- Loss of, or reduced, radio navigation and communications

Icing can form as slowly as ½ inch per hour or as rapidly as 1 inch per minute! Icing potential exists anytime visible moisture—clouds or precipitation—exists at temperatures of 0°C or less. This contradicts the notion that icing can only occur in clouds. We mentioned that freezing rain can be the most serious icing hazard and the icing potential of wet snow. Both phenomena can affect the VFR pilot. The greatest icing potential occurs between the freezing level and −10°C to −15°C, or within a layer approximately 5000 and 7500 feet deep. Icing has been encountered in convective clouds at altitudes of 30,000 to 40,000 feet in temperatures less than −40°C. During the winter of 1996–1997 central California was struck by a series of storms called the "Pineapple Express." These storms began in the latitudes of Hawaii and brought moist, unstable air to California. Icing potential was enhanced by the upslope from the Sierra Nevada mountains. On several occasions aircraft were reporting moderate to severe mixed icing up to FL 260. The temperature at this altitude was −26°C.

Icing is classified by its formation and appearance. Pilots should use these terms when reporting icing.

- Rime icing
- Clear icing
- Mixed icing

Rime Ice

Rime ice is milky, opaque, and granular. It is normally formed when small supercooled water droplets instantaneously freeze upon impact with the aircraft. It is most frequently encountered in stratiform clouds at temperatures between 0° and −20°C. This icing condition is usually widespread due to the character of stratiform clouds. Rime ice is relatively easy to remove with ice protection equipment.

Clear Ice

Clear ice is glossy and formed when large, supercooled water droplets flow over the aircraft's surface after impact and freeze into a smooth sheet of solid ice. It is most frequently encountered in cumuliform clouds or freezing precipitation. Brief but severe accumulations occur at temperatures between 0° and −10°C, with reduced intensities at lower temperatures, and in cumulonimbus clouds down to temperatures as low as −25°C. Clear ice is usually not as widespread as rime ice because it occurs in cumuliform clouds, but its intensity tends to be more severe. Clear ice is more difficult to remove than rime ice.

Rime Ice and Clear Ice (Mixed Ice)

Mixed ice is a hard, rough, irregular, whitish conglomerate formed when supercooled water droplets vary in size or are mixed with snow, ice pellets, or small hail. Deposits become blunt, with rough bulges building out against the airflow. Of the three types of ice, mixed ice is the most difficult to remove.

A condition favorable for rapid accumulation of clear ice is freezing rain below a frontal surface. Icing can also become severe in cumulonimbus clouds along a surface cold front or above a warm front. Icing is also more probable and more severe in mountainous regions than over flat terrain. Mountain ranges cause rapid upward vertical current on the windward side, which supports large water drops. The movement of frontal systems across mountain ranges often combines frontal lift with upslope effect, creating extremely severe icing zones, with the most severe icing taking place above the crest and to the windward side of the ridges.

As with turbulence, pilots have a tendency to overestimate icing intensity, especially when new or low-time. A recently rated instrument pilot, after experiencing his second encounter with icing in a Cessna 172, reported the intensity as severe. The encounter lasted about 30 minutes. The pilot was unable to maintain altitude and forced to descend. This description, however, is only of moderate intensity. Icing intensity has been classified for reporting purposes in the *Aeronautical Information Manual* (AIM) and *Aviation Weather Services*. However, as with turbulence, I prefer more descriptive definitions:

trace Ice becomes perceptible and the rate of accumulation is slightly greater than the rate of sublimation. It is not hazardous even though ice protection equipment is not used, unless encountered for more than one hour. Your spouse admires how pretty it looks on the wing.

light The rate of accumulation can create a problem if the flight continues for more than one hour. Occasional use of ice protection equipment removes or prevents accumulation. Ice should not present a problem if the ice protection equipment is used. Your student hasn't noticed the ice yet; your pilot friend in the back seat is hoping he has enough life insurance; you're negotiating with ATC for a lower altitude, which they can approve in 15 miles. It only take about eight minutes, but each minute seems like 10.

moderate The rate of accumulation, even for short periods, becomes potentially hazardous, and the use of ice protection equipment or flight course diversion becomes necessary. On his second encounter with ice, a friend and his passengers, in an aircraft without ice protection equipment, survived moderate icing only because the terrain was lower than the freezing level.

severe The rate of accumulation is such that ice protection equipment fails to reduce or control the hazard. Immediate diversion is necessary. In this situation, the person in the left seat very rapidly ceases being the pilot and becomes a passenger; the wing is an ice cube. Certain pilots report icing intensity as heavy. This is a misnomer—all ice is heavy!

As a result of several accidents due to icing, the Aviation Weather Center (AWC) of the National Weather Service has revised its icing advisories—AIRMETs and SIGMETs—to imply hazards due to *supercooled large droplet,* or SLD, occurrences.

SLDs are freezing drizzle or freezing rain-sized supercooled water droplets, as opposed to cloud droplets, which are much smaller. A forecast for SLDs implies rapid accumulation of mixed or clear ice, possibly forming aft of aircraft ice-protected areas.

To alert pilots of potential SLD occurrences aloft, the AWC issues advisories containing the terms MIXED OR CLEAR ICING IN CLOUDS OR PRECIPITATION (MXD/CLR ICICIP) or CLEAR ICING IN PRECIPITATION (CLR ICGIP), which means precipitation-size drops aloft. Even though the rate of accumulation is only moderate, the presence of SLDs poses a significant hazard, even to aircraft with ice protection equipment.

To indicate areas of SLD, there may be an AIRMET within an AIRMET to highlight the threat. For example, an AIRMET for moderate rime icing below 14,000 feet may cover a relatively large area. A second AIRMET wholly within the first AIRMET's coverage may forecast moderate mixed or clear icing in clouds and precipitation below 10,000 feet. This AIRMET alerts pilots to the SLD threat within the second area below an altitude of 10,000 feet, with a potential of cloud-size droplets between 10,000 and 14,000 feet.

It appears the most likely areas for SLD occurrence are

- 25 to 300 miles ahead of a warm front
- 30 miles on either side of an occluded front
- 25 to 130 miles ahead of a Pacific cold front
- 25 to 130 miles behind an arctic front

The term *ice protection equipment* in the icing intensity definitions refers to aircraft and equipment certified for flight in known icing conditions. Although many aircraft have limited ice protection equipment (pitot heat, prop anti-ice, alternate static source, etc.), it should never be construed as allowable for flight in icing; their purpose is for emergency use only, should icing be inadvertently encountered.

This brings up the question: What is known icing? You won't find it in FAR Part 1, *Definitions and Abbreviations,* or the Pilot/Controller Glossary. Icing is difficulty to forecast and transitory in nature. Do we want a forecast of icing to forbid flight? Would a report of light icing above 8500 feet, with bases 8000 and tops 9000 preclude flight for aircraft not certified for flight in known icing? What if the terrain was at 7800 and tops 15,000? Do we really want a hard answer? If some in the FAA had their way, the definition would be any time there is visible moisture and a temperature of 5°C or less! Should this or a similar proposal ever be adopted, it would certainly mark the end to many useful icing pireps. If icing is reported or forecast and we fall out of the sky and survive or require emergency or special handling from ATC, we're a candidate for a violation.

A National Transportation Safety Board (NTSB) decision in 1993 held a pilot in violation of Federal Aviation Regulations. The NTSB found that a pilot cannot pick and choose between forecasts and pireps. A forecast for icing is sufficient to warrant the violation.

More often than not, however, ATC is so busy and so happy to get us out of its hair we'll never hear another word. No one should interpret this as meaning that I, the FAA, or ATC condones such actions. Icing for aircraft not certified for flight in icing conditions is to be avoided! At present, the decision as to whether flights can be made safely rests solely with the pilot, which is where it will stay until we prove we're not worthy of the responsibility.

Pireps are the only source of reported icing. Forecasts for icing are contained in the AIRMET Bulletin and SIGMETs.

Structural Icing

Structural icing accidents accounted only for about 40% of total accidents involving icing. The majority of icing accidents are attributed to carburetor or induction system icing, with less than 10% involving icy runways. Both induction and runway icing are addressed later in this chapter. Most structural icing accidents occurred when the pilot continued flight into known icing, severe weather, or deteriorating weather conditions. A lesser amount occurred on approach or landing in icing conditions or with ice accumulation.

Another hazard of structural icing is tailplane or empennage stall, as well as wing and fuselage icing. A tailplane stall occurs, like a wing stall, when the critical angle of attack is exceeded. Since the horizontal stabilizer counters the natural nose-down tendency caused by the center of lift of the main wing, the airplane reacts by pitching down, sometimes uncontrollably, when the tailplane stalls. Application of flaps can aggravate or initiate the stall. A pilot should use caution when applying flaps during an approach if there is the possibility of icing on the tailplane.

Perhaps the most important characteristic of a tailplane stall is the relatively high airspeed at the onset and, if it occurs, the suddenness and magnitude of the nose-down pitch. A stall is more likely to occur when the flaps are approaching the fully extended position, after nose-down pitch and airspeed changes following flaps extension, or during flight through gusty winds.

Another type of icing that must not be ignored is *frost.* As you recall, condensation is the change of state from water vapor to liquid water. When moist air comes in contact with a cool surface and then cools to its dewpoint, dew appears. If the dewpoint temperature is below freezing, the water vapor sublimates—changes from a vapor to a solid—producing frost.

The effects of frost may be more subtle than ice. Although the airfoil's aerodynamic contour remains relatively unchanged, considerable roughness, resulting in increased drag, occurs. Under no circumstances should a takeoff be attempted with frost on the aircraft. A heavy coating of frost can cause a 5% to 10% increase in stall speed. Just as there is no such thing as a little pregnant, there is no such thing as a little frost.

We made a trip from Mammoth Lakes to Livermore, California, in late June. As is my habit, I planned an early morning departure, assuming density altitude the most significant factor. Arriving at the airport around dawn, to my surprise I found frost on the wings! I pointed the airplane into the sun, and in about 15 minutes, we wiped the melting frost from the ship. This just goes to prove you have to be ready for, and aware of, everything.

Induction or Carburetor Icing

As mentioned earlier, the largest number of icing accidents are attributed to induction or carburetor icing. Most such accidents involved the lack, or improper use, of carburetor heat.

On a flight from Van Nuys, California, to San Francisco in a Cessna 172, we encountered light icing after an ATC instruction to climb. I periodically applied carburetor heat. Something unusual occurred. With the carburetor heat on, the engine ran fine; with it off, the engine faltered. On the ramp at San Francisco, we parked next to a Navion that had also flown from L.A., but at a higher altitude, encountering more ice. Sure enough, in the Navion's air filter was a large chunk of ice. I realized that the carburetor heat in the 172 was functioning as an alternate air source. I'm sure this seems ridiculously obvious; it didn't at the time, which illustrates the hazards of learning by experience.

The induction system includes the air filter, ducting, and fuel metering device. Induction-system icing consists of any ice accumulation that blocks any component of the system.

Air-filter icing occurs when flying in areas of visible moisture with temperatures at or below freezing. For VFR pilots, air-filter icing should only occur in areas of freezing precipitation or wet snow.

Induction-system icing takes place anytime structural icing occurs. A symptom of air-filter icing is a more-or-less gradual decrease in power. Should air-filter icing occur, apply carburetor (carb) heat or alternate air. The application of carb heat or alternate air bypasses the air filter. Leave carb heat or alternate air on until above-freezing temperatures melt the ice. The use of carb heat or alternate air results in unfiltered air entering the induction system. Except for operational checks, avoid engaging either control on the ground.

In addition to air intake icing, normally aspirated engines can develop ice in the carburetor throat. Refer to Fig. 4-15. The vaporization of fuel, along with the adiabatic expansion of air as it passes through the fuel discharge nozzle, venturi, throttle valve, and passages to the engine, causes sudden and significant cooling. If the air temperature drops below the dewpoint, water vapor in the air condenses into water droplets. Therefore, water can form in the carburetor in cloudless skies. This cooling can reduce the temperature in the carburetor to below freezing, and, with sufficient moisture present, ice will form. Known as carburetor icing, ice can form with outside air temperatures as high as 32°C. The formation of carburetor ice restricts engine power and may result in complete engine stoppage.

As air accelerates through the carburetor and fuel evaporates, temperatures can be lowered as much as 34°C. Whether ice develops depends on the velocity of the fuel/air mixture, outside air temperature, humidity, and carburetor system. Conditions most favorable for carburetor ice are outside temperatures between −8° and 15°C, high relative humidity, and low power settings.

Carburetor heat preheats the air before it reaches the carburetor. Carburetor heat is usually adequate to prevent icing, but it may not always clear ice that has already formed. Pilots should monitor engine performance for the first signs of icing, especially during favorable conditions. When ice is detected or suspected, immediately apply full carburetor heat. Leave heat on until all ice has been removed.

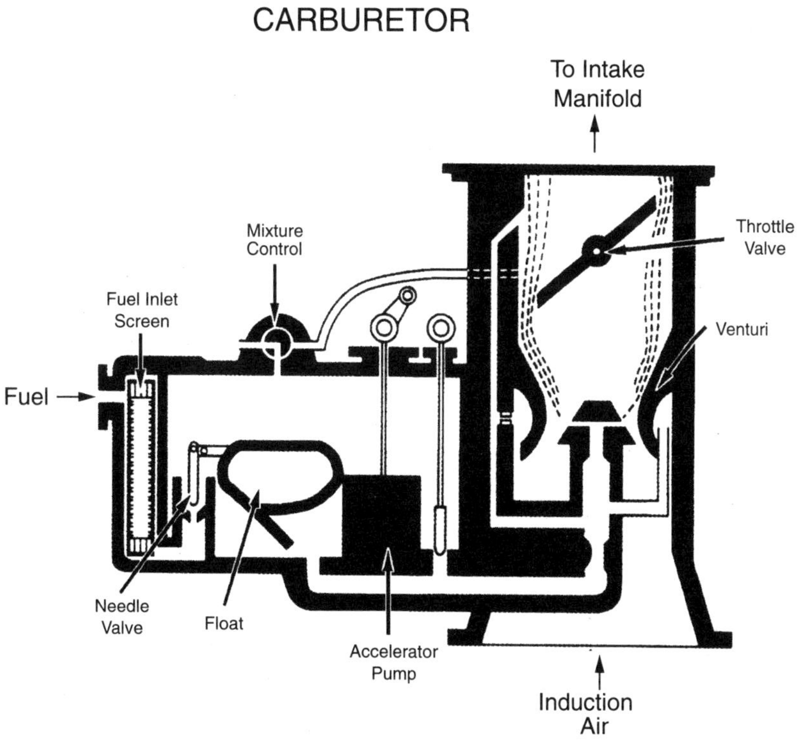

Fig. 4-15. *The vaporization of fuel, along with the adiabatic expansion of air as it passes through the carburetor, can result in temperatures below 0°C.*

Using full carburetor heat initially causes an additional loss of power and engine roughness. The added loss of power and roughness results from the richer mixture due to warmer, less-dense air and melting ice passing through the engine. You must resist the natural urge to remove carb heat. Leave full heat on until an increase in rpm or manifold pressure and smooth engine operation resume.

Under severe conditions, it may be necessary to leave carburetor heat on for an extended period. When carb heat must be left on, relean the mixture for maximum rpm and smoothest operation. A cruise power setting of 75% or less with any amount of heat will not damage the engine.

On a flight from Page, Arizona, to Las Vegas, Nevada, in our Cessna 150, conditions for carburetor ice were ideal. The temperature was about 15°C in rain showers. At the first indication of carb ice, heat was applied. It cleared up the ice, but as soon as it was removed, rpm began dropping. The only solution was to leave carb heat on and live with the hundred or so loss in engine rpm, even after releaning the mixture.

Carbureted engines are more susceptible to icing during reduced-power operation. Some aircraft and engine manufacturers recommend the use of carburetor heat during all power reductions; others recommend it only when ice is suspected. Pilots should know and follow the aircraft manufacturer's recommendations. If full power is required, such as in a go-around, full carburetor heat and full power might cause early detonation or engine damage. It certainly prevents the engine from developing full power, which might be critical in low-power aircraft at high-density altitudes. Again, know and follow the manufacturer's recommendations.

Pay attention! I had remained overnight in Amarillo, Texas, because of a line of thunderstorms that approached from the west. The Cessna 150 was parked into the wind when torrential rains moved through the area. The next morning was clear, with abundant surface moisture, and the temperature was about 15°C, with nearly 100% relative humidity.

My first clue of trouble was the increased throttle setting required to obtain idle rpm. Engine runup also took more throttle than usual. I suspected carburetor ice and a water-saturated air filter because of the conditions. I had 13,000 feet of runway.

Full throttle only gave me about 2200 rpm. The increased ground run to rotation speed—about 7000 feet—should have been another clue. I was off the ground, and with no runway remaining and 200 feet of altitude, the engine started losing rpm. I applied carburetor heat and the engine was running very rough, producing about 1700 rpm.

There was a tremendous psychological urge to reduce carb heat and get that rpm back. I was preparing to crash straight ahead, but the engine was still producing power and I decided to make a 180° turn and land on a taxiway. Then I informed a surprised tower controller of what happened; remember, a pilot's first job is to fly the airplane.

After running the engine for 20 minutes, and one aborted takeoff later, I launched into the air. The engine again performed normally above the shallow, moist layer. It was a perfect example of having the clues and ignoring them. I was extremely fortunate.

Some airplanes have a carburetor air temperature gauge. This gauge measures the temperature in the carburetor. The yellow arc indicates temperatures at which icing is most likely to occur. When ice is suspected or detected, the pilot can apply enough heat to raise the temperature out of the danger zone. Do not use partial carburetor heat without a carburetor air temperature gauge. Applying partial heat or leaving it on for an insufficient time might aggravate the situation.

STRATEGIES

If the weather briefing indicates even a remote possibility of encountering ice, the pilot should accomplish several tasks. Ensure that the pitot heat works with a very light touch during preflight. (Be careful, it gets hot!) Check ice protection equipment for proper operation. Remember that a heated pitot is an anti-icing device, to be turned on before encountering ice. Deicing equipment usually requires ice buildup before activation; improper operation can actually increase ice buildup and prevent its removal! Check alternate air or carburetor heat for an alternate source of air should the air filter ice over.

The pilot of a twin-engine airplane was en route when the airplane encountered icing conditions. When the pilot activated the deice boots, the right wind deice boot failed to function. As the airplane slowed for landing, the asymmetrical ice buildup caused instability that the pilot was unable to control. The airplane crashed on landing, seriously injuring the occupants. There was no mention of the pilot checking the boots prior to departure.

One of our local pilots at Fresno had a similar occurrence in a Piper Aerostar. During the first storm of the season, one deicing boot inflated upon activation the other did not. Our pilot was visibly shaken by the experience. The bottom line: Check all aircraft equipment prior to departure.

As a new instrument instructor, I took an instrument student on a flight from Van Nuys to Lancaster's Fox Field. The freezing level was forecast to be at 6000 feet. The minimum altitude for the route was 7000. Cloud bases were at 6500, well above terrain. Sure enough, we picked up trace to light rime icing. We had neglected to turn on the pitot heat and a reverse cone of rime ice grew from the pitot tube. I matter-of-factly pointed this out to my student, and we turned on the heat, which immediately corrected the problem. On descent into Fox, the ice made a deafening nose as it broke off the tail surfaces. I do not recommend this procedure!

The object with ice is to minimize exposure. With temperatures 0°C or less, avoid flying in clouds or precipitation. Should ice be encountered, immediately notify ATC and initiate a plan of action. The first consideration, if the aircraft has sufficient performance, might be to climb to colder air or above the clouds based upon pireps and the weather briefing. Ice slowly sublimates—changes from a solid directly into a gas—when on top. Or, descend to warmer air based upon the actual freezing level on climbout. You did note the freezing level on climbout? This requires a careful check of minimum en route altitudes (MEA). Finally, if you have to, turn around; presumably you came from an ice-free area. The point is, do something!

A nonturbocharged Baron without ice protection equipment departed Reno, Nevada, for southern California. Moderate icing and severe turbulence were forecast. The pilot elected to fly a direct course along the crest of the Sierra Nevada mountains, the route where the most intense icing and turbulence could be expected. The aircraft iced up, resulting in a fatal accident.

The pilot had no way out because the MEA was the aircraft service ceiling. The terrain was well above the freezing level, and the pilot failed to reverse course at the first sign of ice. What other options were available? The pilot could have crossed the mountains near Sacramento, minimizing exposure to ice, and once over the Sierras, it was all downhill. The pilot could have flown toward Las Vegas where the weather was considerably better or simply waited for better weather conditions.

When the Baron became ice-covered, the pilot had no option but to ride it to the crash site. Attempting flight under these conditions and with this type of equipment was quite literally suicide.

A Bonanza pilot departed the San Francisco Bay area on a flight to Los Angeles. Icing above 7000 feet was forecast and reported. The pilot elected to fly at 11,000 feet. The pilot's last words were, "I've iced up and stalled." The crash occurred in the San Joaquin Valley, where the elevation was near sea level. Minimum altitudes in the vicinity of the crash were well below the freezing level. The pilot simply did nothing until aircraft control was lost.

An advantage of turbocharged and pressurized aircraft is the ability to fly high, above the weather. Icing is normally not a significant factor in the flight levels, except around convective activity or in the summer when temperatures can range between 0°C and −10°C at these altitudes. Just because it's summer doesn't necessarily mean there is no icing potential.

The following story illustrates the decision-making process with forecast icing. The aircraft was a turbo Mooney on a flight from Bakersfield to Hayward, California. The synopsis indicated moisture and a stable air mass. Bases along the route were reported around 5000 and tops 9000 to 11,000 feet, the freezing level 7000. Except for the coastal mountains, terrain along the route was close to sea level. Terrain—as we have mentioned—is a very important factor. The flight was planned at 12,000 feet, because tops were relatively low, and the aircraft had the performance to quickly climb through the potential icing layer. During the climb, trace to light icing was encountered. Once on top, the ice sublimated quickly. There were some buildups above 12,000. Deviations to avoid these clouds were obtained from ATC. By circumnavigating the buildups, icing and turbulence were avoided. You can put money on the assumption that there was ice in those clouds. Should this be attempted in an airplane with lesser performance? Absolutely not! The bases and tops were known quantities. The airplane had the performance to quickly climb on top. Had that not been possible, the pilot had the option to return. Cloud bases were more than 4000 feet above terrain and well below the freezing level. Was the icing forecast correct? Yes. Had emergency assistance been required, the pilot would have been a candidate for a violation. I do not intend to imply that this procedure is recommended or endorsed. The decision rests solely with the pilot, based on his or her training and experience, and the capability of the aircraft.

If a descent through an icing layer is required, remember the objective is to minimize exposure. Under such circumstances, negotiate with ATC to obtain a continuous descent. Avoid, if possible, level flight in clouds. ATC is usually very responsive to such requests.

A popular notion in some aviation publications is that a pilot's mere mention of ice will receive emergencylike handling. Icing might be an emergency, but remember the controller's job is to separate aircraft within a finite amount of airspace. ATC might have to assign a higher altitude, but ATC cannot, and should not, be expected to fly the aircraft or assume the responsibility of pilot in command. To paraphrase: An accurate report of actual icing conditions is worth a thousand forecasts.

In freezing precipitation, aircraft without a heated pitot and alternate static source, especially in IFR conditions, would be in serious trouble. Another significant factor, especially for aircraft without ice protection equipment, is that accumulated ice could be carried all the way to the ground, making landing extremely hazardous. It cannot be overemphasized that this hazard can affect VFR as well as IFR operations. Should this phenomena be encountered in aircraft without ice protection equipment, virtually the only option, and certainly the safest, is to fly into warmer air and land. Pilots who fly into ice with aircraft not certified for flight in icing conditions must have the right stuff, because, in every sense of the word, they become test pilots.

Continually updating the weather picture is the key to managing a flight, especially in aircraft with limited or no ice protection equipment. Icing can be significant during descent, especially when the destination temperature is at or below freezing. Flight Watch can provide information on tops, temperatures aloft, reported and forecast icing, and current surface conditions.

On airplanes equipped with a float-type carburetor, the use of carb heat is normally recommended during reduced (out of the airspeed indicator's green arc) or closed throttle operations. Since the heat is generated by the running engine, during extended periods of low throttle operation, it may be desirable to advance the throttle periodically to ensure proper engine operation. On airplanes with a pressure carburetor, which is much less susceptible to icing, carb heat may only be recommended if icing conditions exist.

Remove carburetor heat prior to a go-around or balked landing. With carburetor heat, loss of power may be critical at low altitude and low airspeed, and engine damage may occur with full heat on a go-around at takeoff power on some powerplants.

Low temperatures cannot be ignored. Snow, snow melt, freezing rain, and frost produce structural ice that can be difficult to remove from a parked aircraft. Even a thin layer of frost can severely affect performance and must be removed before takeoff, according to federal regulations.

If an aircraft is parked in an area of blowing snow, special attention should be given to openings in the aircraft where snow can enter, freeze solid, and obstruct operations. These openings should be free of snow and ice before flight. Some of these areas are

- Pitot tubes
- Static system sensing ports
- Wheel wells
- Heater intakes
- Carburetor intakes
- Tail wheel area
- Control surfaces and cables

Some years ago, four of us flew a Cessna 172 into the South Lake Tahoe Airport in November. After our stay, about ¾ inch of ice had formed on the airplane. Being naive at the time about such conditions, I assumed it would blow off during the takeoff roll—silly me.

An experienced tower controller, however, suggested we clean the ice. It had to be scraped off with a plastic scraper! Later I calculated we had between 300 and 400 pounds of ice on the airplane. I had no experience with this condition at the time. I shudder to think what would have happened during a takeoff in a low-performance airplane, 300 pounds over gross, at a high-altitude airport, with ¾ inch of ice on the airfoils. This is a perfect example of where the test almost came before the lesson.

This was the case with a Cessna Caravan pilot. The aircraft was parked outside when freezing rain fell. The next day the pilot removed about 80% of the snow covering the wings, which left a coarse layer of ice about 3/16 inch thick. The aircraft crashed after takeoff. The probable cause was determined to be the pilot's failure to remove ice from the airframe prior to takeoff.

The solution to these problems is to hangar the aircraft, arrange for a deicing service, or plan the departure later in the day when the sun has melted the ice. During taxi, avoid areas of standing water or slush. Slush thrown into wheel wells, wheel pants, and control surfaces can freeze, resulting in locked controls and frozen landing gear or brakes. With a descent through an icing layer and surface temperatures close to or below freezing, a pilot must be prepared for an approach and landing with airframe icing, possibly to an ice- or snow-covered runway.

If snow, ice, or slush are on the runway, aircraft control might be difficult, especially in high winds, with reduced braking efficiency, resulting in a longer-than-normal ground roll. Without windscreen deice—that's deice, not defrost—a pilot could be faced (pardon the pun) with zero forward visibility during the landing. Regulations require the pilot to consider, "...runway lengths at airports of intended use...." Pilots operating in this environment must consider these factors when selecting destination and alternate airports or even the advisability of making the flight.

During the winter, both VFR and IFR pilots may have to contend with a snow- or ice-covered runway. Airports may be closed due to snow or ice accumulation or for snow removal. Be sure to check Notices to Airmen (NOTAMs) for airport closures, runway conditions, and braking action reports. It's also a good idea to check en route with an FSS for updates on destination and alternate runway conditions.

Civil pilots now have access to objective runway friction measurements at some airports. The Greek letter *mu* (μ), pronounced "myôô," designates a friction value representing runway surface conditions. Values range from 0 to 100, where 0 is lowest friction value and 100 is the maximum. With snow or ice on the runway, a mu value of 40 or less is the level when the aircraft braking performance starts to deteriorate and directional control begins to be less responsive. The lower the mu value, the less effective braking performance becomes and the more difficult direction control becomes.

Airport management conducts friction measurements on runways covered with snow or ice. Values of 40 or less are reported for the first third, middle third, and last third of the runway. Pilots can expect to receive mu values from ATC or NOTAMs.

No correlation has been established between mu values and the descriptive terms "good," "fair," "poor," and "nil" used in braking action reports. Pilots should use mu information with other knowledge, including type, weight, wind conditions, and previous experience. The bottom line: If you haven't had training or experience with snow- or ice-covered runways, avoid them!

Arriving pilots during instrument weather conditions may have to descend through an icing layer all the way to the ground. On landing a pilot may have an iced-over windscreen and adverse flight characteristics, including tailplane icing. When the pilot lowers the flaps for landing, the change in angle of attack might cause the tailplane to stall. This is why some people advise the use of no flaps when ice is present. Higher stall speed due to ice may require a higher approach airspeed. With ice, no flaps, and a higher-than-normal airspeed, a longer-than-normal ground roll, possibly on an ice-covered runway, may be required.

Recall our discussion of flight and instrument icing problems. The solution to an iced-over pitot or engine instrument probe is crosscheck. Crosschecking all flight or engine instruments reveals the discrepancy. With the problem diagnosed, a plan can be developed to work around the situation.

The investigation of a Cessna 152 accident revealed that the crankcase breather line was plugged with ice, and the oil had been forced out through the engine's nose seal. The probable cause was determined to be the pilot's inadequate preflight that failed to detect the ice-clogged breather.

As with turbulence, there are no hard and fast rules as to exactly where ice will develop. I had planned to include a table providing an overview of icing locations, probability, and intensity—similar to the table in Chapter 3 on turbulence. However, after consulting with NWS icing experts, it seems that such generalities are just not possible. This fact highlights the hazards associated with icing.

Winter can be one of the best flying seasons, with its cool temperatures and usually excellent flying weather between storms. But, like mountain flying in the summer, winter flying has its own unique problems and solutions.

5
Families of the air

IN CHAPTER 1 WE DISCUSSED THE STRUCTURE OF THE ATMOSPHERE and in Chapter 3, general circulation. Like much of our discussion, we build on the knowledge gained in those chapters and this chapter to better understand weather theory and phenomena and its application to flight planning and inflight decision-making.

This chapter begins with a discussion of air masses, their source regions, and properties. From there we move on to air mass modification, then to air mass influence on aviation weather.

Within our discussion of air masses, we present the effects of low and high pressure areas. Pressure patterns indicate areas at the surface under the influence of high or low pressure, troughs, or ridges. The terms high and low, like hot and cold, are relative. A high is simply an area completely surrounded by lower pressure. Conversely, a low is an area surrounded by higher pressure. Convergence occurs in areas of low pressure, divergence in areas of high pressure. Surface convergence and divergence only affect the lower 10,000 feet of the atmosphere but can be major factors in the weather machine, creating vertical motion. In general, areas of low pressure are associated with poor weather, and high pressure, good weather, although we will find that there are no absolutes and a number of exceptions to this general rule.

According to Peter E. Kraght in *Airline Weather Services 1931-1981,*

> A disastrous thunderstorm accident close to Bowling Green, Kentucky, in 1943 that involved an American (Airlines) DC-3 started a chain of events that

> eventually led to the first systematic research into thunderstorm behavior. The plane crashed onto the ground either near or under a severe thunderstorm. Buell (C.E. Buell, chief meteorologist, American Airlines 1939-1946) initiated a letter to the Civil Aeronautics Board pointing out the appalling dearth of understanding of what actually occurs inside a thunderstorm, as evidenced by the accident investigation. He recommended a massive research effort be organized to probe into thunderstorms and document their internal structure.

Thunderstorms were and are a major hazard to aviation operations. It has been said that sooner or later, a pilot must contend with thunderstorms. In this section we define and dissect this phenomenon and related hazardous weather.

As aircraft instruments and navigational system capabilities improved, pilots began taking on more and more weather. Hazards of fog and low clouds were solved with the instrument landing system (and increased fuel reserves). Icing, to a large extent, was overcome in the 1930s and, for the most part, turbulence and thunderstorms were mastered with high cruise altitudes and radar. With these hazards virtually solved, major air carrier accidents most often fall into the categories of mechanical failure, pilot error, and windshear caused by microbursts. (Typically, major air carrier accidents are caused by a combination of the previous causal factors, any one of which would not have resulted in an accident, but in combination was disastrous.)

Microbursts are inherently random and difficult to forecast. The hazards of windshear microbursts affect general aviation as well as air carrier operations. In this portion of the chapter we discuss the causes, recognition, and avoidance of this phenomena.

In the final section of the chapter, "Strategies," we evaluate accident statistics. With a knowledge of thunderstorm phenomena, we can develop techniques to avoid their hazards. The section concludes with some "dos" and don'ts" of thunderstorm avoidance and recommended strategies should a thunderstorm be inadvertently penetrated. However, it cannot be overemphasized that the key to dealing with thunderstorm hazards is avoidance!

AIR MASSES

An air mass is a widespread body of air with homogeneous—that is, similar—properties. Air masses take on the properties of the region where they originate. These areas are called source regions. Air masses have similar temperature and moisture content through both their horizontal and vertical cross sections. As air masses migrate, they undergo modifications to their temperature, moisture, and lapse rate.

To better understand air mass source regions, it might be helpful to review Fig. 3-1. As discussed in Chapter 3, at the pole is the polar high, which is stable. Between the polar easterlies and the midlatitude westerlies is an area of global convergence that accounts for the polar front, which is unstable. At around 30° latitude are the subtropical highs, an area of surface divergence, which is stable. In the vicinity of the equator is the intertropical convergence zone, another area of surface convergence, which is unstable.

When air masses develop and remain relatively stationary in their source regions, they take on the characteristics (temperature, moisture, and stability) of that region. Air mass source regions are arctic, polar, and tropical, which reflect their temperature. To indicate moisture content air masses are classified as continental (dry) or maritime (moist). The following air masses source regions affect North America:

- Polar ice cap
- Plains of northern Canada
- North Pacific Ocean
- North Atlantic Ocean
- Gulf of Mexico
- Southwestern United States
- Northern Mexico

Air stagnating over northern continental regions forms continental polar or continental arctic air masses, depending on temperature. (Remember from Chapter 1 that temperatures are relative. Cold means colder than the underlying surface; warm means warmer than the underlying surface.) These air masses are characterized by cold, dry air reflecting their source region and are very stable. Maritime polar air masses form over northern oceanic areas and are generally not as cold as continental polar, especially in winter. Why? As discussed in Chapter 1, water surfaces tend to retain heat, while land areas cool rapidly. Maritime polar air masses tend to contain more moisture than their continental counterparts and are conditionally unstable. Maritime tropical air masses form over warm oceanic areas in lower latitudes and are very moist and usually unstable. Arid continental regions produce continental tropical air masses, which are hot, dry, and unstable.

Refer to Fig. 5-1. Continental polar, and sometimes arctic, air masses develop over Canada and Alaska. They push down through Canada into the United States, sometimes reaching as far south as California and the Gulf of Mexico—cold, dry, and stable. The North Pacific and North Atlantic oceans spawn maritime polar air masses—cold, moist, and conditionally unstable. Since the general flow over midlatitudes is from west to east, the United States is rarely affected by maritime polar air masses that develop over the North Atlantic. The west coast, Gulf coast, and Atlantic coast are affected by maritime tropical air masses—warm, moist, and unstable. Finally, from northern Mexico and the desert southwestern United States come continental tropical air masses—warm, dry, and unstable.

Dust devils, or dust whirls, form in continental tropical air masses. This microscale whirlwind forms over hot, dry land in fair weather and light winds as the result of intense local daytime convection and surface friction. They have diameters of 10 to 50 feet and extend from the surface to several thousand feet. Windspeed within the rotations vary from 25 to more than 75 knots. Dust devils are capable of substantial damage, but the majority are small. I was doing pattern work with a student at Lancaster's Fox Field in the California Mojave Desert when we flew into a dust devil. The encounter was equivalent to light-to-moderate turbulence; it shook the Cessna 150, and the low pressure in the vortex caused both windows to pop open! Dust devils most likely form

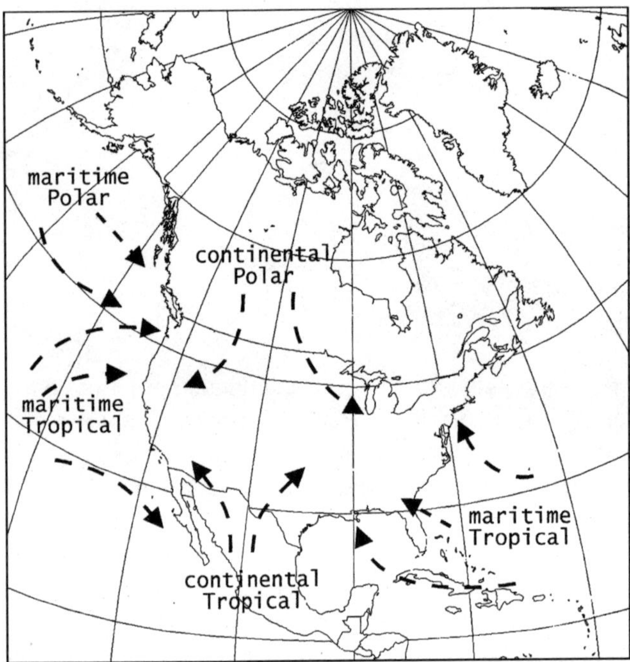

Fig. 5-1. *An air mass is a widespread body of air with homogeneous properties.*

in the same manner as fair-weather waterspouts (discussed later in this chapter), but, lacking the energy released from latent heat, they are smaller.

When an air mass moves over an area of different properties, the air mass takes on some or all of the surface properties. Horizontal changes in temperature, moisture, and lapse rate gradually change. For example, let's take a maritime air mass that moves over the continent. In winter the air mass is cooled from below. Its lapse rate becomes more stable. In the summer the air mass is warmed from below, resulting in greater instability. These effects play an important part in our discussion of fronts in Chapter 6. When an air mass moves over a colder surface, the term "cold" is added to the type. For example, when our maritime air mass moves over the continent, it is termed a "maritime polar cold" air mass. In the summer if it moves over warmer land it would be called a "maritime polar warm" air mass. Table 5-1 compares the general characteristics of air masses. Like Tables 2-1 and 4-1, there are many similarities.

Figure 5-2 shows a cross section of a low and a high pressure area. The central pressure in the low is 800 mbs, in the high 1000 mbs. Constant pressure surfaces (850, 700, and 500 mbs) are at different heights within each air mass. Recall our discussion of flying from "high to low, look out below" in Chapter 1. From Fig. 5-2 we can easily see that without correcting the altimeter, the aircraft would indeed be flying at a lower true altitude as it approached the low pressure center.

Low Pressure Areas and Troughs

A low is an area completely surrounded by higher pressure. A *trough* consists of an elongated area of low pressure, almost always associated with cyclonic wind flow (counterclockwise in the northern hemisphere). In common practice, the terms *cyclone* and *low* are used interchangeably. Both terms refer to a closed circulation; a cyclone or low is not a trough.

Convergence—upward vertical motion—occurs in lows and troughs. Convergence destabilizes the atmosphere, which increases relative humidity and clouds. Convergence itself does not necessarily produce poor weather. Other factors, such as moisture and stability, must be considered. For example, a surface low or trough may only mix—stir up—the lower atmosphere. Without moisture and instability, this mixing would produce good visibility and possible low-level turbulence. However, a surface low or trough with a moist, unstable air mass may be all that's needed to trigger thunderstorms.

Table 5-1. Characteristics of Air Masses

Warm—cooled from below	**Cold—warmed from below**
Stable lapse rate	Unstable lapse rate
Poor visibility	Good visibility
Smooth	Turbulent
Stratiform clouds	Cumuliform clouds
Steady precipitation	Showery precipitation
Rime icing	Clear icing

SURFACE PRESSURE

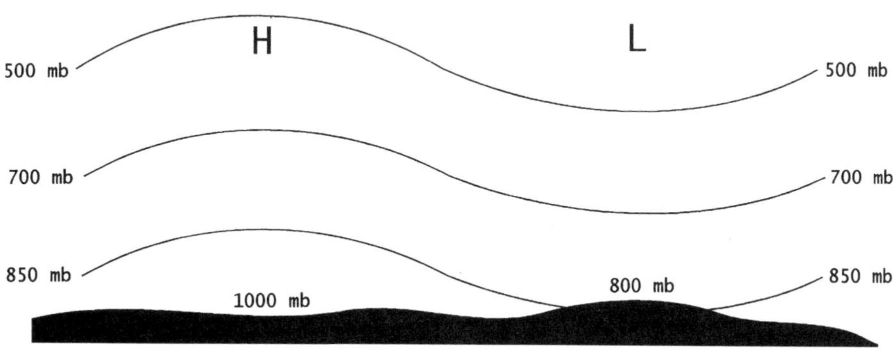

Fig. 5-2. *Without correcting the altimeter, the aircraft would indeed be flying at a lower true altitude as it approaches the low pressure center.*

Surface convergence occurs along curved isobars surrounding a low or trough. Maximum convergence takes place at the center of the low or along the trough line.

Troughs are not fronts, although fronts normally lie in troughs. A front, as we shall see in the next chapter, is the boundary between air masses of different temperatures, whereas a trough is simply a line of low pressure. Both phenomena produce upward vertical motion.

A thermal low is an area of low atmospheric pressure caused by high temperatures produced by intensive heating of the earth's surface. Thermal lows are common to the continental subtropics in summer. They remain stationary over the area. Cyclonic circulation is generally weak and diffuse. They are not associated with fronts.

Thermal lows develop over the desert southwest in late spring through early fall. At times they can extend to the arid regions east of the Cascade Mountains in Washington and Oregon and the Snake River Valley of Idaho. Their primary effect is to lower pressures inland, increasing coastal marine layers, and they provide thermal convection and weak convergence for the development of air mass thunderstorms.

High Pressure Areas and Ridges

A high is an area completely surrounded by lower pressure. A ridge, the opposite of a trough, is an elongated area of high pressure, almost always associated with anticyclonic wind flow (clockwise in the northern hemisphere). Like cyclone, *anticyclone* refers to a closed circulation and is used interchangeably with high.

Divergence—downward vertical motion—occurs in highs and ridges. Divergence stabilizes the atmosphere, which decreases relative humidity and clouds. Divergence does not necessarily produce good weather.

There is a misconception that high pressure always means good flying weather. Although good weather often occurs, there are exceptions. Strong pressure gradients at the edge of high cells can cause vigorous winds and severe turbulence. Near the center of a high, or with weak gradients, moisture and pollutants can be trapped at lower levels, causing reduced visibility and even producing zero-zero conditions in fog for days or even weeks.

A thermal high results from the cooling of air by a cold underlying surface. For this cooling to occur, the air mass must remain relatively stationary over the cold surface, a factor in the development of continental polar and arctic air masses.

THUNDERSTORMS

A thunderstorm is a local storm produced by cumulonimbus clouds. The storm itself may be a single cumulonimbus cloud or cell, a cluster of cells, or a line of cumulonimbus clouds that in some cases may extend for several hundred miles. Thunderstorms, as the name implies, are always accompanied by lightning and thunder. These storms usually produce strong, gusty winds, heavy rain, and sometimes hail. Individual cells last for a short period, seldom more than two hours. Lightning is the visible electrical discharge produced by thunderstorms; thunder is the sound produced by rapidly expanding gases along the channel of a lightning discharge.

For manual weather observation, a thunderstorm is reported when thunder is heard or overhead lightning or hail is observed. Lightning (thunderstorm) detectors are not yet available at most automated observation sites. When installed, lightning detectors will report a thunderstorm when detected in the vicinity of the station. As the definition of a thunderstorm in the previous paragraphs indicates, a cumulonimbus cloud implies a thunderstorm. Therefore, any report of thunder, lightning, or cumulonimbus clouds means thunderstorm, with all of its ominous implications.

When surface winds gusting to 50 knots or more or hail $\frac{3}{4}$ inch or greater accompany a thunderstorm, the storm is classified as severe. Before METAR, severe thunderstorms were reported in surface observation. In the METAR code, a severe thunderstorm is implied with winds of 50 knots or more or hail $\frac{3}{4}$ inch or greater. METAR reports contain the intensity of precipitation produced by thunderstorms. It's important to remember that although precipitation and thunderstorms are always associated, they are two separate phenomena. Thunderstorms may be reported with no precipitation or with any intensity of precipitation—light, moderate, or heavy. Some pilots mistranslate reports as a light thundershower or a light thunderstorm and rain showers; neither is correct. Light rain or rain showers may accompany a thunderstorm. However, there is no such thing as a light thunderstorm!

When a thunderstorm exists completely within nonconvective clouds and precipitation, it is termed embedded. Embedded thunderstorms are especially dangerous because they can be impossible to detect without storm detection equipment—radar or lightning detectors.

Three conditions are necessary for thunderstorm formation:

- An initial lifting mechanism
- Sufficient water vapor
- Unstable air

If any one of the three elements is missing, thunderstorms will not develop.

All thunderstorm cells progress through three stages, called the life cycle:

- Cumulus
- Mature
- Dissipating

Thunderstorm life cycle is illustrated in Fig. 5-3.

The first ingredient required for thunderstorm development is an initial lifting mechanism. In Chapter 2, we discussed vertical motion in the atmosphere. Any of those phenomena can produce the required lifting. Most often, lifting is produced by a frontal surface, sloping terrain, convergence, or surface heating. However, upslope and low level warm-air advection may be all that's required to trigger thunderstorms.

This lifting creates an initial updraft. At the lifted condensation level, adiabatic cooling in the updraft results in condensation and the beginning of cloud development. This is the beginning of the cumulus stage.

THUNDERSTORM LIFE-CYCLE

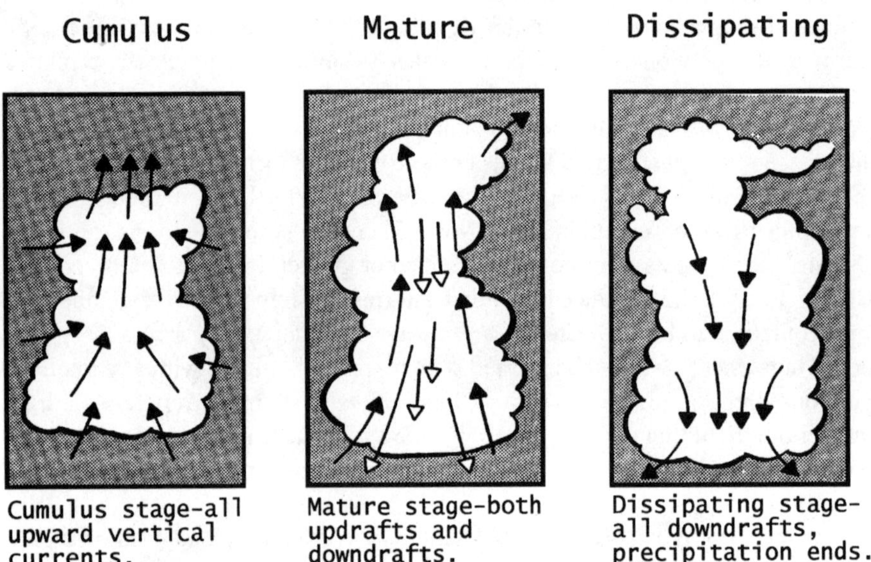

Fig. 5-3. *All thunderstorms progress through the three stages of the life cycle.*

The cumulus stage gives birth to the thunderstorm. Although most cumulus clouds do not become thunderstorms, the initial stage is always a cumulus cloud or collection of cumulus clouds. Figure 5-4 shows the cumulus stage. The main feature of the cumulus or building stage is the predominant updrafts, which may extend from the surface to several thousand feet above the visible cloud top. During the early period of this stage, cloud droplets are very small, but they grow into raindrops as the cloud builds upward.

If we think of the thunderstorm as an engine, the initial lifting creating the updraft is the "starter." "Ignition" is the initial condensation. "Fuel" is the water vapor.

Without sufficient water vapor, the initial lifting may not be sufficient to sustain continued vertical movement, and a thunderstorm will not form. If the atmosphere is stable, vertical motion ceases and stratiform clouds form. However, if the air is unstable—the third requirement for thunderstorm development—cumulus clouds form.

Condensation releases latent heat, which partially offsets the adiabatic cooling in the saturated updraft, increasing buoyancy. That is, the rising air is warmer than the surrounding air. The increased buoyancy drives the updraft still faster, drawing more water vapor into the cloud. The updraft becomes self-sustaining in the unstable air, and a thunderstorm develops. Cold-air advection aloft destabilizes the atmosphere and enhances thunderstorm development.

As the cloud towers upward, the updraft cools adiabatically until in upper levels of the storm it becomes colder than the surrounding air. This difference in temperature, together with the increasing weight of water drops and ice particles, retards the upward motion, ultimately turning it to a downdraft or outflow. The change of flow may directly reverse the updraft, or it may arch outward as an outflow, allowing the updraft to continue unabated. This is the mature stage of the thunderstorm.

In the mature stage, drops are ejected from the updraft or become so large that the updraft can no longer support them and they begin to fall. The mature stage begins when rain first falls from the bottom of the cloud. Updrafts continue to gain strength in the early mature stage, which may exceed 5000 feet per minute.

Figure 5-5 shows the mature stage. This stage occurs roughly 10 to 15 minutes after the cloud has built upward above the freezing level. As the raindrops fall, they pull air downward, which is a major factor in the formation of downdrafts in the mature stage. The air dragged downward by the falling rain cools the surrounding air, which accelerates downward motion. Throughout the mature stage, downdrafts continue to develop and coexist with updrafts. Maximum updrafts tend to be found in the upper two-thirds of the storm. Updrafts as strong as 6000 feet per minute and downdrafts up to 2000 feet per minute can develop.

The mature stage represents the most intense period of the thunderstorm. Lightning activity is greatest, and hail, if present, most often occurs in the mature stage. The essential difference between a nonsevere and severe thunderstorm is the mechanism of the mature stage, which is discussed later in this section.

Fig. 5-4. *The cumulus stage is characterized by a collection of cumulus clouds and updrafts throughout the cloud.*

Fig. 5-5. *The mature stage is characterized by both updrafts and downdrafts.*

Back to our engine analogy: "Energy" to drive the storm is the latent heat released by condensation; the engine or storm now becomes self-sustaining—it no longer depends on the starter or initial lifting. "Combustion" products are clouds and precipitation.

The dissipating stage begins when water vapor is cut off and downdrafts predominate. The dissipating stage is characterized by weak downdrafts throughout the cloud. Figure 5-6 shows the dissipating stage. Condensation gradually decreases, and when all water has fallen from the cloud or has evaporated, the dissipating stage is complete. At the surface, all signs of the thunderstorm disappear and any clouds that remain are stratiform.

A final look at our engine: "Exhaust" is the outflow or downdraft. The throttle is the cutting off of fuel—water vapor. Individual storms vary, particularly in the structure of the exhaust and throttle, a determining factor as to the severity of the storm.

Structure in the mature stage has an important bearing on the duration and potential severity of a storm and leads to two general thunderstorm classification categories: *limited state thunderstorm,* generally referred to as air mass thunderstorms, and *steady state thunderstorms.*

Limited State Thunderstorms

Thunderstorm cells that progress rapidly though the mature stage are usually limited state. The mature stage is self-destructive. When the updraft is too weak to support the rain drops, precipitation falls through the updraft. Falling precipitation induces frictional drag, retarding the updraft and finally reversing it to a downdraft. The downdraft and precipitation cool the lower portion of the storm cloud and the surface below, cutting off the inflow of water vapor and dissipating the storm. This self-destructive cell usually lasts from 20 to about 90 minutes and rarely produces extreme turbulence or large hail. However, even a limited state thunderstorm is capable of producing severe icing and turbulence and hail that can cause structural damage.

In the summer season, air mass thunderstorms that occur west of the Rockies are most frequent over the higher mountains, deserts, and plateaus. Moisture can be brought in from as far away as the Gulf of California or even the Gulf of Mexico. (Fig. 5-7). The anticyclonic flow of high pressure over the Gulf of Mexico has pushed moist, unstable air through Arizona and New Mexico, into Utah and Colorado. This moist air usually spreads north and can even reach as far as Idaho at least two or three times during the season.

Figure 5-7 also shows air mass thunderstorms in the lower Mississippi Valley, Ohio Valley, and southeastern United States. They do not appear as distinct as those in the west because of the time of day. This is a visual satellite image taken at 2331 UTC (approximately 4 P.M. Pacific Daylight Time or 7 P.M. Eastern Daylight Time).

In the western United States, weak fronts, and most importantly the passage of upper-level troughs, help to set off summer thunderstorms. When moist tropical air is lifted

Fig. 5-6. *The dissipating stage is characterized by weak downdrafts throughout the cloud.*

Fig. 5-7. *In the West, the moisture for air mass thunderstorms can come from as far as the Gulf of California or Gulf of Mexico.*

by a cold front, severe thunderstorms can develop. Although infrequent, lines or clusters can occur. They most often develop along California's coastal mountains, the Sierra Nevadas, Cascades, west slopes of the Wasatch Mountains in Utah, the mountains east of Burley and Idaho Falls in Idaho, and the mountains of northeast Nevada. Mountains provide a primary lifting mechanism for the development of these thunderstorms (Fig. 5-8). A line of thunderstorms has developed along the Sierra Nevada mountains and clusters of thunderstorms have developed in northeast Nevada, northwest Utah, and Idaho.

Air mass or single-cell thunderstorms tend to follow the 700-mb wind flow. Therefore, the 700-mb constant pressure chart can be used as a guide for the movement of these storms.

Steady State Thunderstorms

Supercell and squall line thunderstorms are the main types of severe, steady state thunderstorms. The supercell thunderstorm, with its enormous updrafts and downdrafts, is able to maintain itself as a single entity for hours. This type of storm produces tornadoes and large hail. These storms can grow to over 60,000 feet high.

When both updrafts and downdrafts coexist in the mature stage and are about equally balanced—not significantly affecting the other—steady state thunderstorms result. The most obvious consequence is that the mature stage continues in this "steady state" and intensifies, becoming severe with extreme turbulence and large hail. The life of an individual cell may be considerably longer than that of a limited state cell. The severe storm complex, often consisting of many dissipating and developing cells, may last as long as 24 hours and move as far as 1000 miles. The thunderstorm continues in this steady state until affected by some outside influence or until the mechanics of the thunderstorm cell itself change and it becomes self-destructive.

Dissipating groups of thunderstorms can produce outflow boundaries. An outflow boundary is a surface boundary left by the horizontal spreading of thunderstorm-cooled air. The boundary is often the lifting mechanism needed to generate new thunderstorms. Outflow boundaries are often depicted on the surface analysis chart.

When thunderstorms form lines, they are often referred to as a *squall line.* Squall lines can develop along a cold front, but more often they appear a hundred or so miles ahead of the front. Squall lines are caused by air aloft flowing over the cold front and developing into waves, much like mountain waves. Where the wave crests—a lifting mechanism—thunderstorms develop in the moist, unstable air.

Another cause of squall lines occurs in the central plains, especially during the spring. Behind the cold front, cold, dry, continental polar air pushes in from the north. Ahead of

Fig. 5-8. *Air mass thunderstorms can form in lines or clusters, as well as individual cells.*

the front, warm, dry, continental tropical air moves in from the southwest. To the east, warm, moist, maritime tropical air advances from the Gulf. Where the continental tropical and maritime tropical air masses meet, a dry line forms, which produces a squall line. Once the thunderstorms develop, the outflow of cold air along the ground initiates lifting, which generates new, and possibly more severe, storms—an outflow boundary.

When conditions are right, individual thunderstorms can grow into large, organized convective weather systems. These are known as mesoscale convective complexes (MCC). MCCs are large, covering an area the size of several states. The following are characteristics of MCCs:

- Light winds aloft
- High moisture content at low levels
- Low-level warm-air advection
- Late afternoon development
- Persistence all night
- Production of widespread areas of rain or rainshowers
- Production of IFR conditions over wide areas
- Heavy rainshowers and thunderstorms
- Thunderstorms usually circumnavigable with storm detection equipment
- Area tends to remain stationary

Individual thunderstorms within the MCC combine to generate a long-lasting, slow-moving weather system. The thunderstorms feed on themselves, producing widespread precipitation. MCCs can produce widespread severe weather, including high winds, hail, and tornadoes. Figure 5-9 illustrates an MCC. The area covers the Texas panhandle, northcentral Texas, and most of Oklahoma.

Thickness is the vertical depth of a layer in the atmosphere between two pressure surfaces. The NWS produces a 1000- to-500-millibar thickness chart. MCCs, unlike air mass thunderstorms, tend to move parallel to the thickness lines on the 1000- to-500-millibar thickness chart.

Large areas of the Midwest are plagued by night thunderstorms. Nighttime variations in the large-scale wind system create convergence in low levels. These thunderstorms show a peak occurrence between midnight and 4 a.m. and are not the remnants of evening thunderstorms left over from daytime convection.

Thunderstorm Hazards

Thunderstorm hazards consist of turbulence, icing, precipitation—including hail—lightning, tornadoes, gusty surface winds—including low-level windshear—effects on the altimeter, and low ceilings and visibility. Under certain conditions, they can produce high-density altitude. Thunderstorm hazards are illustrated in Fig. 5-10.

It should be no surprise that thunderstorms have the potential to produce severe to extreme turbulence. Vertical motion is the structural basis of the cell. In thunder-

Fig. 5-9. *Individual thunderstorms can grow into a large, organized convective weather systems, known as a mesoscale convective complexes (MCCs).*

storms, the width of up- and downdrafts may vary from a few feet to several thousand feet. These drafts affect the aircraft's altitude as it flies through the thunderstorm. It is virtually impossible to hold altitude. Altitude changes of several thousand feet are not unusual. This is illustrated by the close proximity of the up- and downdrafts in Fig. 5-10.

Downdrafts continue below the base of the cloud with significant speed to within 300 to 400 feet of the ground. These drafts constitute a significant hazard to flight beneath the thunderstorm, which is often in heavy rain and poor visibility. The most severe turbulence is frequently encountered near the freezing level but can also occur from the ground to above the cloud tops. Significant turbulence can also occur in clear air well away the cell itself.

Icing is another principal hazard in thunderstorms. Expect icing in all storms at elevations above the freezing level. Although thunderstorm clouds are usually limited in diameter, even short duration can result in severe icing. The most severe icing can be expected between the freezing level and −15°C, as illustrated in Fig. 5-10.

Thunderstorms contain great amounts of liquid moisture, even though rain might not be falling. The amount of liquid water decreases above the freezing level, where snow becomes more predominant. Snow mixed with supercooled water exists everywhere above the freezing level, creating an icing hazard.

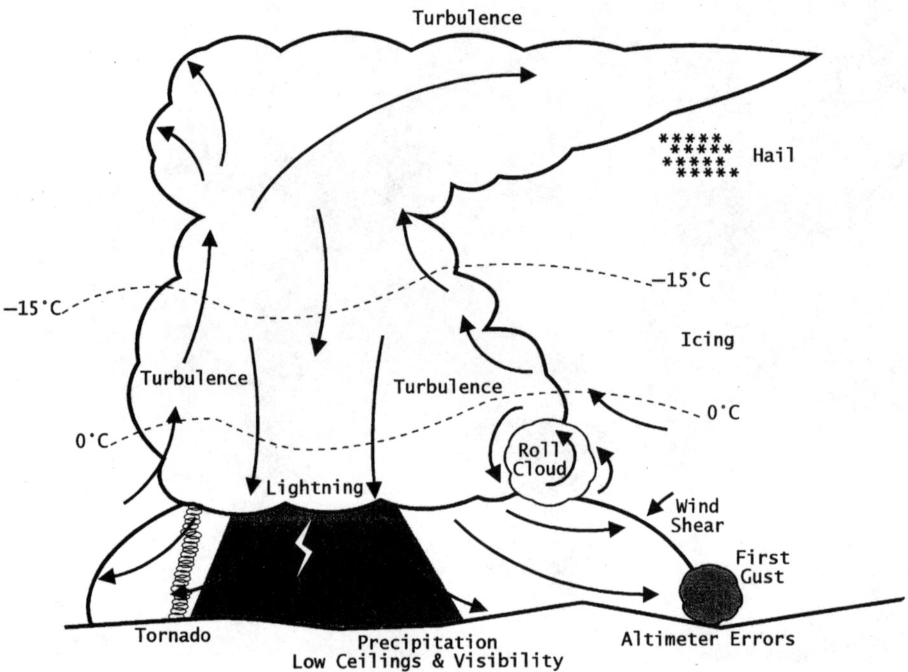

Fig. 5-10. *Thunderstorms can produce every kind of aviation weather hazard.*

Hail can be one of the worst hazards of thunderstorm flying. Great amounts of hail and the largest stones generally are found in the larger and taller storms, but many thunderstorms have no hail associated with them. In general, large hail occurs in severe thunderstorms. Hail batters airfoils, particularly the leading edges, and in extreme cases has knocked out windscreens and caused engines to fail. Frequently hail is carried aloft and tossed out the top or side of the cloud by updrafts and may be encountered in clear air several miles from the cloud. Hail frequently exists in thunderstorms even though not reported at the ground. Flight beneath the anvil should be avoided because of the hail hazard, as illustrated in Fig. 5-10.

Lightning experienced in a thunderstorm can cause temporary blindness so that control of the aircraft by reference to instruments may be momentarily lost. Damage to navigational and electronic equipment by lightning also can create a hazard. Small punctures in aircraft skin may result from direct lightning strikes. Lightning is found throughout the thunderstorm cloud but is most frequent and severe from the freezing level up to −10°C. The aircraft, like cars on the ground, typically insulates the passengers from lightning hazards.

The only known incident of lightning downing a jetliner occurred on December 8, 1963. Over Elkton, Maryland, lightning struck an aircraft, exploding three of its fuel tanks. Eighty-one people perished. Although lightning strikes an airplane approximately every 3000 hours, significant damage is the exception, rather than the rule. Another exception occurred in May 1996 when lightning struck a Beech King Air. The result was a cabin fire. Lightning strikes can burn wire, magnetize airframes, destroy composite structures, fuse control surfaces, cause turbojet compressor stall and flameout, and, although very rare, ignite fuel tanks. The answer: Avoid thunderstorms!

Lightning is classified by its origin and destination. Lightning can be from cloud to ground, in cloud, cloud to cloud, cloud to air. Astronauts have observed lightning coming out of the tops of clouds, cloud to space. Figure 5-11 shows lightning that would classified as cloud to ground.

Another electrical phenomenon associated with thunderstorms is corona discharge, colloquially known as St. Elmo's fire. St. Elmo's fire becomes visible as bluish static electric streaks dancing across the windscreen. Aircraft flying through or in the vicinity of thunderstorms often develop corona discharge streamers from antennas and propellers and even from the entire fuselage and wing structure. It produces the so-called precipitation static. Precipitation static, however, usually only affects low-frequency radio communications and navigation.

Tornadoes are produced by severe thunderstorms. They are whirlpools of air, cloud, and debris that range in diameter from 100 feet to a ½ mile. Pressure is extremely low in the center of the small, concentrated vortex. Tornado winds probably reach 200 to 300 knots, although, based on damage patterns, winds of more than 400 knots are indicated. Tornadoes appear as funnel-shaped clouds from the base of thunderstorms and usually

Fig. 5-11. *The solution to lightning hazards is to avoid thunderstorms.*

move at 25 to 50 knots. Their paths range from a few miles to probably less than 50 miles, although squall lines and frontal systems can produce a series of tornadoes that can cover hundreds of miles. Their exact path is erratic and unpredictable.

Technically, they must touch the ground to be called tornadoes. When they occur over water, they are known as waterspouts. When the characteristic whirling clouds extend downward from the parent cloud but do not reach the surface, they are called funnel clouds. Families of tornadoes or funnels have been observed from appendages of a main cloud extending several miles outward from the area of lightning and precipitation. A tornado vortex extends a great distance into the parent cloud and pilots may encounter extreme turbulence in the imbedded vortex, which is not visible. Frequently, cumulonimbus mamma clouds occur in connection with violent thunderstorms and tornadoes.

An average of approximately 200 tornadoes per year occur in the United States; about 90% develop ahead of cold fronts. Tornadoes are not a common factor for the west coast states or the intermountain region of Idaho, Montana, Nevada, Utah, Arizona, and the western portions of Wyoming, Colorado, and New Mexico. Tornadoes occur most frequently in the Great Plains states east of the Rocky Mountains. However, they have occurred in every state and Canada. Tornadoes occur with isolated thunderstorms at times but more frequently with cold fronts or squall line thunderstorms. Reports or forecasts of tornadoes are indications that atmospheric conditions in the area are favorable for extreme turbulence.

The most severe thunderstorms develop in a windshear environment. That is, windspeed must increase with height. At the same time, the wind direction veers (a clockwise change in wind direction). Typically, surface wind blows from the southeast at between 15 and 20 knots; at 5000 feet wind is from the south at between 30 and 35 knots; at 15,000 feet wind is from the southwest at about 50 knots. This allows the storms to have separate updrafts and downdrafts—steady state thunderstorms. Low-level air feeds the storm from the southeast. This air is lifted in an updraft and exits the storm at high levels toward the east. Air at mid-levels approaches the storm from the southwest. It passes around the updraft, gets caught in a downdraft on the north side of the storm and exits on the back side as part of the low-level outflow. Pilots can expect to see discussions of this windshear environment in the outlook portion of convective SIGMETs and convective outlooks.

Figure 5-12 illustrates the typical synoptic situation for tornadoes in the central United States. A low-level jet of moist air feeds the systems from the south. The upper-level tropopause jet acts like a vacuum sucking up the moist air. The dashed-line box shows the usual location where tornadoes develop.

Forecasts for tornadoes are contained in convective SIGMETs, alert weather watches, severe weather watch bulletins, and the convective outlook.

Fair-weather waterspouts form over warm and shallow coastal waters. They are much smaller and less intense than the average tornado and tend to form through convergence in unstable air beneath developing cumulus clouds. They tend to be stronger than their dust devil cousins because of the energy available, through the release of latent heat, from the ocean surface.

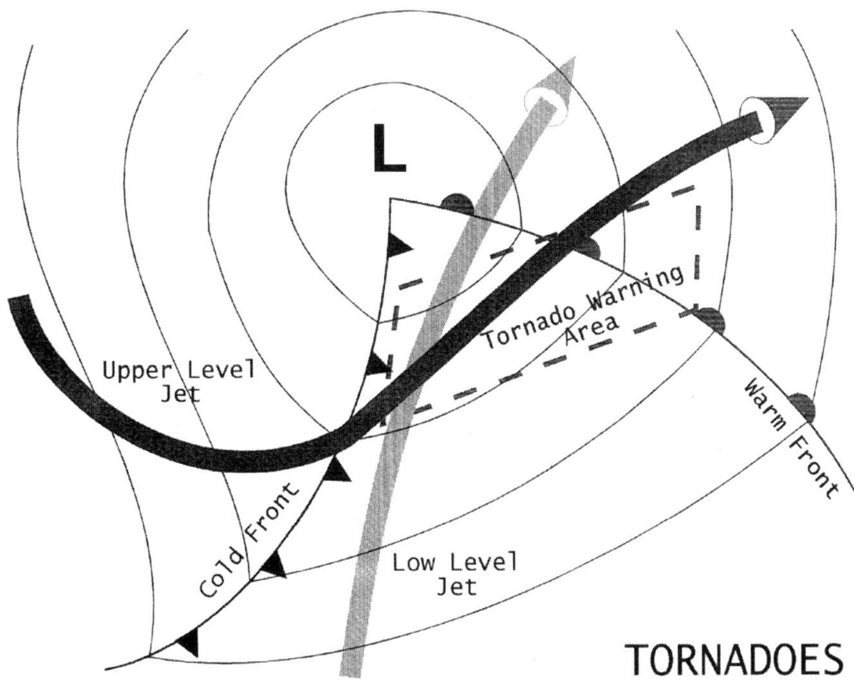

Fig. 5-12. *The typical synoptic situation for tornadoes in the central United States.*

Refer to Fig. 5-10. Gusty and variable surface winds are associated with thunderstorms. Usually the first gust, or gust front, precedes the arrival of the cloud and onset of rain as the thunderstorm approaches. Frequently it stirs up dust and debris as it plows along, announcing the thunderstorm's approach. The strength of the first gust frequently is the strongest observed at the surface during a thunderstorm. It may approach 100 knots in extreme cases. The roll cloud is not always present, but it is found most frequently on the leading edge of fast-moving thunderstorms. It represents an extremely turbulent area.

A *gustnado* describes a funnel cloud that develops along the gust front; the funnel cloud is not a tornado. It is believed that the gustnado receives its initial rotation from the shift in wind directions across the gust front. Cold, dense air behind the gust front lifting the warm air ahead imparts a rotating motion in the windshear zone.

Pilots should avoid takeoff or landing when a thunderstorm is within 10 to 20 miles of the airport. This range is the region of the strongest and most variable winds. Caution must also be exercised following thunderstorm passage. A strong, gusty outflow boundary can follow the storm. Along with these winds are downbursts and microbursts, which produce severe low-level windshear. Since microbursts and severe low-level windshear are such hazardous phenomena, they are covered in a separate section.

Pressure usually falls rapidly with the approach of a thunderstorm. It then rises sharply with onset of the gust front and arrival of the cold downdraft and heavy rain. Pressure then normally falls back as the storm moves on. This cycle of pressure change may occur within 15 minutes. Height indicated on a pressure altimeter during the storm may be in error by more than a hundred feet.

Heavy rain brings lowering ceilings and visibility. With severe storms, ceilings, visibility can be at or near zero. Figure 5-13 shows a thunderstorm downburst in Idaho's Snake River Valley.

We have already mentioned the thunderstorm phenomenon of a heat burst. Carolyn Kloth, forecaster at the NWS's Aviation Weather Center, has proposed that a heat burst may be a new thunderstorm hazard for aviation—high-density altitude. A heat burst increases turbulence and windshear, develops in radar echo-free air, and affects the pressure altimeter. They typically occur in the dissipating stage of nighttime thunderstorms.

CONVECTIVE LOW-LEVEL WINDSHEAR

In August 1985, a Delta Air Lines L-1011 crashed at the Dallas/Fort Worth Airport. The NTSB was unable to determine if the crew had been using airborne weather radar at the time of the crash. The NTSB report did state, however, "The evidence concerning the use of the airborne weather radar at close range was contradictory. Testimony was offered that the airborne weather radar was not useful at low altitudes and in close proximity to a weather cell…" although, "At least three aircraft scanned the storm at very close range near the time of the accident." The accident was probably caused by a microburst from a single severe storm cell, which illustrates how weather can develop rapidly, often without any severe weather warning.

Fig. 5-13. *With severe storms, ceilings and visibility can be at or near zero.*

Windshear, a rapid change in wind direction or speed, has always been around. Convective activity produces severe shear, which is defined as a rapid change in wind direction or velocity causing airspeed changes greater than 15 knots or vertical speed changes greater than 500 feet per minute. The microburst produces the most severe windshear threat.

Rain-cooled air within a thunderstorm produces a concentrated rain or virga shaft less than ½ mile in diameter, which forms a downdraft. The downdraft or downburst has a very sharp edge and forms a ring vortex upon contact with the ground. It spreads out, causing gust fronts that are particularly hazardous to aircraft during takeoff, approach, and landing. Reaching the ground, the burst continues as an expanding outflow.

A microburst consists of a small-scale, severe, storm downburst less than 2½ miles across. This flow can be 180° from the prevailing wind, with an average peak intensity of about 45 knots. Microburst winds intensify for about five minutes after ground contact and typically dissipate about 10 to 20 minutes later. Microburst windspeed differences of almost 100 knots have been measured. On August 1, 1983, at Andrews Air Force Base, indicated differences near 200 knots were observed. Some microburst events are beyond the capability of any aircraft and pilot to recover. Although they are normally midafternoon, midsummer events, microbursts can occur any time, in any season.

Avoidance is the best defense against a microburst encounter. When the possibility of microbursts exists, the pilot must continually check all clues. Therefore, pilots must learn to recognize situations favorable to this phenomenon.

Figure 5-14 illustrates the effects of a microburst on a landing aircraft. This is the same scenario that claimed the Delta flight. The aircraft is established on the glidepath and may even have the runway in sight. The outflow from the microburst causes increasing headwind, which has the effect of causing the aircraft to climb. Typically the pilot increases the rate of descent. Just as the rate of descent increases, the aircraft flies into a strong downdraft, increasing the rate of descent. The pilot attempts to climb back to the proper glidepath, which may result in reduced airspeed just as the aircraft enters the area of increasing tailwind. If the aircraft is close to the ground, recovery may not be possible.

The scenario is similar for aircraft taking off. The aircraft encounters increased headwind during the takeoff roll. At rotation or just after liftoff, the aircraft encounters the strong downdraft, immediately followed by an increasing tailwind. Again, if the aircraft is close to the ground, recovery may not be possible.

Windshear Recognition

The following discussion is based on AC 00-54, *Pilot Windshear Guide* and a Department of Commerce publication, *Microbursts: A Handbook for Visual Identification.* The latter publication is for sale by the Superintendent of Documents. It contains an in-depth, technical explanation of the phenomena along with numerous color photographs depicting microburst activity. It should be part of every pilot's library.

Microbursts can develop any time convective activity, such as thunderstorms, rain showers, or virga, occur, associated with both heavy and light precipitation. Approximately 5% of all thunderstorms produce microbursts, and more than one microburst can

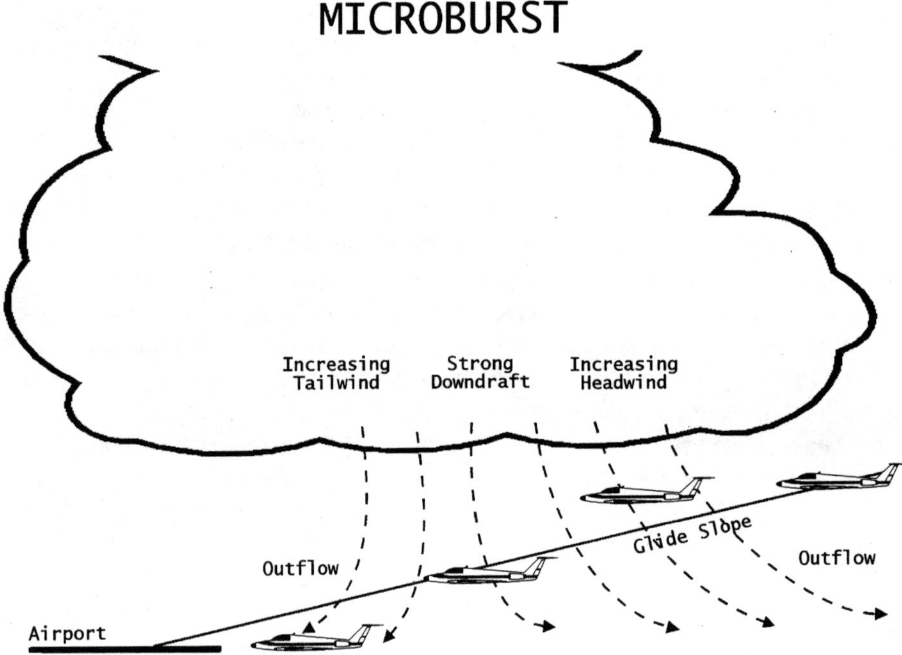

Fig. 5-14. *Avoidance is the best defense against a microburst encounter.*

occur with the same weather system. Therefore, pilots must be alert for additional microbursts if one has already been encountered or reported and prepare for turbulence and shear as subsequent microbursts interact. Microbursts are characterized by precipitation or dust curls carried back up toward the cloud base, horizontal bulging near the surface in a rain shaft, forming a foot-shaped prominence, an increase in windspeed as the microburst expands over the ground, and abrupt wind gusts.

Microbursts can occur in extremely dry as well as wet environments. The lack of low clouds does not guarantee the absence of shear. Microbursts can develop below clouds with bases as high as 15,000 feet. As virga or light rain falls, intense cooling causes the cold air to plunge, resulting in a dry microburst. Evaporative cooling turbulence associated with this phenomena has already been discussed. Anvils of large dryline thunderstorms can produce high-level virga and result in dry microbursts. High-based thunderstorms with heavy rain should be of particular concern. This was the type that produced intense windshear in the 1985 Delta accident in Dallas/Fort Worth.

Embedded microbursts are produced by heavy rain from low-based clouds in a wet environment. A wet microburst might first appear as a darkened mass of rain within a light rain shaft. As the microburst moves out along the surface, a characteristic upward curl appears.

The potential for windshear and microbursts exists whenever convective activity occurs. Pilots should review forecasts for thunderstorms—thunderstorms imply low-level windshear. Check surface reports and pireps for windshear clues: thunderstorms, rain showers, gusty winds, or blowing dust. Dry microbursts are more difficult to recognize. Check surface reports for convective activity (rain showers, cumulonimbus clouds, and virga) and low relative humidity (15° to 30° temperature/dewpoint spread).

The low-level windshear alert system (LLWAS) has been installed at 110 airports in the United States. The system detects differences between windspeed around the airport and a reference center-field station. Differences trigger an alert. Sensors are not necessarily associated with specific runways; therefore, descriptions of remote sites are based on the eight points of the compass: "Center field wind three one zero at one five. North boundary wind zero niner zero at three five."

The *Airport/Facility Directory* advertises the availability of LLWAS under "Weather Data Sources." The lack of a LLWAS alert does not necessarily indicate the absence of windshear. LLWAS has limitations. The magnitude of the shear might be underestimated. Surface obstructions can disrupt or limit the airflow near the sensor, and, due to location of sensors, microburst development might go undetected, especially in the early stages. Sensors are located at the surface; therefore, microburst development that has not yet reached the surface will be undetected, and because coverage only exists near the runways, microbursts on approach will not be observed. Even with these limitations, LLWAS can provide useful information about winds near the airport.

Development continues on an automated terminal Doppler weather radar (TDWR) system that is based on the same principle as NEXRAD. TDWR will provide windshear and microburst warnings to controllers and should become operational in the late 1990s.

Airborne radar returns of heavy precipitation indicate the possibility of microbursts. Although potentially hazardous, dry microbursts might only produce weak radar returns. Strong windshear might occur as far as 15 miles from storm echoes. Radar echoes can be misleading by themselves, and it might require a Doppler radar to spot the danger of a dry microburst. The southwest edge of an intense storm can appear weak both visually and on radar; however, this area is known to spawn tornadoes and severe windshear. Convective weather approaching an airport on the downwind side tends to be more hazardous than activity moving away.

No quantitative means exists for determining the presence or intensity of microburst windshear. Pilots must exercise extreme caution when determining a course of action. Microburst windshear probability guidelines have been developed by the FAA and apply to operations within 3 miles of the airport, along the intended flight path, and below 1000 feet agl. Probabilities are cumulative; therefore when more than one point exists, probability increases.

The following factors indicate a high probability of windshear with the presence of convective weather near the intended flight path.

- Reports of localized strong winds or observations of blowing dust, rings of dust, or tornado-like features
- Visual or radar indications of heavy precipitation
- Pireps of airspeed changes of 15 knots or greater
- LLWAS alert or wind velocity changes of 20 knots or greater

A pilot must give critical attention to these observations. A decision to avoid, divert, or delay is wise.

The following factors indicate a medium probability of windshear with the presence of convective weather near the intended flight path.

- Rain showers, lightning, virga, or moderate or greater turbulence reported or indicated on radar.
- A temperature/dewpoint spread of 15°C to 30°C.
- Pireps of airspeed changes less than 15 knots.
- LLWAS alert or wind velocity changes of less than 20 knots.

A pilot should consider avoiding these conditions. Precautions are indicated.

The FAA states: "Pilots are…urged to exercise caution when determining a course of action." Probability guidelines "…should not replace sound judgment in making avoidance decisions." In aviation weather, there are no guarantees. The lack of high or medium probability indicators in no way promises the absence of windshear when convective weather is present or forecast. Avoidance is the best precaution.

Takeoff, Approach, and Landing Precautions

Select the longest suitable runway for takeoff. Determine at what point the takeoff can be aborted with enough runway to stop the aircraft. Certain manufacturers provide tables for takeoff calculations; otherwise, the pilot has to base this distance on landing roll tables and experience. Use the recommended flap setting for gusty wind or turbulent conditions, if available. Use maximum-rated takeoff power to reduces takeoff roll and overrun exposure. Consider increased airspeed at rotation to perhaps improve the ability of the aircraft to negotiate windshear or turbulence after liftoff. Do not use a speed reference flight director. Be alert for airspeed fluctuations that might be the first signs of windshear. Should shear be encountered with sufficient runway remaining, abort the takeoff. This decision, however, can only be made by the pilot, based on training and experience. After takeoff, use maximum-rated power and rate of climb to achieve a safe altitude, at least 1000 feet agl.

Select the longest suitable runway to land. Consider a recommended approach configuration with a higher-than-normal approach speed. Turbulent air penetration or maneuvering speed should be considered. Establish a stabilized approach at least 1000 feet agl with configuration, power, and trim set to follow the glideslope without additional changes. Any deviation from glideslope or airspeed change indicates shear. The autopilot, except for autoflight systems, should be disengaged, with the pilot closely monitoring vertical speed, altimeter, and glideslope displacement. Groundspeed and airspeed comparisons can provide additional information for windshear recognition. Increased approach speed, while providing an extra margin for safety, requires longer than normal landing distance.

Windshear Recovery Technique

Windshear recovery technique has not yet been developed for small aircraft. The following windshear recovery technique, developed for airline aircraft, has been adapted from

AC 00-54. It is, however, logical and applicable to most windshear encounters in practically any aircraft.

Windshear recognition is crucial to making a timely recovery decision. Encounters occur infrequently with only a few seconds to initiate a successful recovery. The objective is to keep the aircraft flying as long as possible in hopes of exiting the shear. The first priority must be to maintain aircraft control. The following guidelines were developed for the airlines; exact criteria cannot be established. Whenever these parameters are exceeded, recovery or abandoning the takeoff or approach should be strongly considered. It must be emphasized that it is the responsibility of the pilot to assess the situation and use sound judgment in determining the safest course of action. It might be necessary to initiate recovery before any of these parameters are reached.

- Plus or minus 15 knots indicated airspeed
- Plus or minus 500 feet per minute vertical speed
- Plus or minus one dot glideslope displacement
- Unusual throttle position for a significant period of time.

If any condition is encountered, aggressively apply maximum-rated power. Avoid engine overboost unless required to avoid ground contact. While on approach, do not attempt to land. Establish maximum rate of climb airspeed. As with any turbulent condition, pitch up in a smooth, steady manner. Should ground contact be imminent, pitch up to best-angle-of-climb airspeed, being careful not to stall the aircraft. Controlled contact with the ground is preferable to an uncontrolled encounter. When aircraft safety has been ensured, adjust power to maintain specified limits. When the aircraft is climbing and ground contact is no longer an immediate concern, cautiously reduce pitch to desired airspeed.

The key to the thunderstorm and LLWS hazard is avoidance. A superior pilot uses superior knowledge to avoid having to use superior skill. At the first sign of severe shear, reject the takeoff or abandon the approach. It is easier to explain an aborted takeoff or missed approach to passengers rather than explain an accident to the FAA and insurance company—assuming you're still around to do so.

STRATEGIES

Accident statistics show the majority of thunderstorm accidents occur to noninstrument-rated, low-time private pilots; over half resulted in fatalities. Ironically, most pilots received a preflight weather briefing. Most accidents occurred when pilots initiated IFR flight into adverse weather, attempted VFR flight into deteriorating weather, or attempted to fly in or around thunderstorms. Some occurred when flight was continued into areas of embedded thunderstorms. Others resulted in loss of control due to high, gusty winds or crosswinds.

The violent nature of thunderstorms causes gust fronts, strong updrafts and downdrafts, and windshear in clear air adjacent to the storm out to 20 miles with severe storms and squall lines. Precipitation, which is detected by radar, generally occurs in the downdraft, while updrafts remain relatively precipitation-free. Clear air or lack of radar echoes does not guarantee a smooth flight in the vicinity of thunderstorms.

Our first defense is a complete preflight weather briefing. Recall our balloon pilots in the Napa Valley. They typically call the FSS and request an abbreviated briefing of specific weather reports and winds aloft forecasts. Since they rarely ask for the synopsis or area forecast, and often scattered thunderstorms are not covered in weather advisories, they would not receive information on convective activity. A complete or standard briefing just prior to departure cannot be overstressed.

Storm detection equipment is airborne weather radar or lightning detection equipment. Both radar and lightning detectors have limitations. Pilots using this equipment to avoid thunderstorms must understand their operation and limitations. Just reading the manual is certainly not enough to prepare a pilot to translate the complex symbology presented into reliable information. Pilots should obtain training courses with appropriate instructors and simulators to properly use this equipment.

More and more general aviation aircraft are equipped with airborne weather radar and lightning detection equipment. However, these systems are plagued by low power, attenuation, and limited range. A pilot might pick his or her way through a convective area only to find additional activity beyond. The following pirep illustrates just such an occurrence:

> MSY UUA /OV NEW 150020/TM 2015/FLDURD/TP C550/TB SEV/RM OCCURRED IN AREA WHERE ACFT RADAR DID NOT INDICATE PCPN. BOTH CREW INJURED.

This incident occurred over New Orleans in thunderstorm weather. Both crew of a Cessna Citation were injured when the aircraft encountered severe turbulence in an area where their airborne weather radar indicated no precipitation.

Lightning detection equipment, tradenamed Stormscope, was invented in the mid-1970s by Paul A. Ryan as a low-cost alternative to radar. The Stormscope senses and displays electrical discharges in approximate range and azimuth to the aircraft. Like radar, Stormscope has limitations. One misconception proclaims that in the absence of dots or lighted bands there are no thunderstorms. However, NASA's tests of the Stormscope differed. Precipitation intensity levels of three and occasionally four (VIP levels) would be indicated on radar without activating the lightning detection system. A clear display only indicates the absence of electrical discharges. This absence does not necessarily mean convective activity and associated thunderstorm hazards are not present. Even tornadic storms have been found that produced very little lightning. The lack of electrical activity, as with the absence of a precipitation display on radar, does not necessarily translate into a smooth ride.

Many authorities agree that a combination of radar and Stormscope is the best thunderstorm detection system. It cannot be overemphasized that these are avoidance, not penetration, devices. Thunderstorms imply severe or greater turbulence and neither radar nor Stormscope, at the present, directly detect turbulence.

Airborne weather radars are low power, generally with a wave length of 3 centimeters. Precipitation attenuation, which is directly related to wave length and power, can be a significant factor. Precipitation attenuation results from radar energy being absorbed and scattered by close targets, and the display becomes unreliable in close proximity to heavy rain or hail. Intensity might be greater than displayed, with distant targets obscured. An accumulation of ice on the radome causes additional distortion.

According to the NTSB, precipitation attenuation was a contributing factor in the crashes of a Southern Airways DC-9 in 1977 and an Air Wisconsin Metroliner in 1980. Precipitation attenuation is not significant with NWS's NEXRAD Doppler units; however, it can be a serious problem with units of 5 centimeters or less, especially in heavy rain. The NTSB recommends:

> ...in the terminal area, comparison of ground returns to weather echoes is a useful technique to identify when attenuation is occurring. Tilt the antenna down and observe ground returns around the radar echo. With very heavy intervening rain, ground returns behind the echo will not be present. This area lacking ground returns is referred to as a shadow and may indicate a larger area of precipitation than is shown on the indicator. Areas of shadowing should be avoided.

One solution to the limitations of airborne weather radar and lightning detection systems is Flight Watch. With real-time NWS NEXRAD Doppler weather radar, Flight Watch has the latest information. Well before engaging any convective activity, a pilot should consult Flight Watch to determine the extent of the system, its movement, intensity, and intensity trend. Armed with this information, the pilot can determine whether to attempt to penetrate the system or select a suitable alternate. ATC prefers issuing alternate clearances compared to handling emergencies in congested airspace and severe weather.

A Bonanza pilot approached an area of thunderstorms in California's Central Valley. The pilot received the latest weather radar and satellite information, as well as pireps and surface observations from Flight Watch. The pilot safely traversed the area with minimum diversion or delay.

This is not a very exciting story, but that's the purpose of Flight Watch, to assist pilots in conducting uneventful flights. En Route Flight Advisory Service has been around for more than 20 years. In spite of this, its function, and the best way to use this important service, is not known to, or is misunderstood by, many pilots.

We were trying to get from Springfield, Illinois, to Oklahoma City. As we arrived in St. Louis, the ground-based weather radar showed cell after cell along our route to the southwest. We decided to wait it out until the next day. However, about six that evening the front passed and the weather cleared. We decided to continue on to Kansas City, which was also behind the front. As darkness fell, the southern horizon was ablaze with continuous lightning.

Checking weather from Kansas City to Oklahoma City indicated clearing, but continuing thunderstorm activity. After departure, we again saw lightning on the horizon. That night it just wasn't meant to be. We returned to Kansas City and spent the night. The next day was bright and clear. I was attending the FAA's En Route Air Traffic Control school at the time. I missed half a day. It would have sure been embarrassing to have been involved in an aircraft accident!

Returning from Oshkosh, thunderstorms were forecast in the Colorado area. Sure enough as we approached Pueblo, there were several cells. A check with Flight Watch indicated Pueblo was in the clear and Colorado Springs would make a suitable alternate, should Pueblo's weather deteriorate. By keeping visual contact with the storms, we circumnavigated to the north around the heavy rain, lightning, and buildups and made an uneventful landing. There really is no reason for getting caught in a cell.

A notion held by some pilots is that the absence of a weather advisory means no significant weather. The lack of a weather advisory does not guarantee the absence of hazardous weather. An unfortunate pilot learned this lesson the hard way. The synopsis described a moist, unstable air mass. Thunderstorms were not forecast for the time of flight, but they were expected to develop; thunderstorms, however, were already being reported along the route. The pilot, without storm detection equipment, encountered extreme turbulence inadvertently entering a cell. The pilot had filed an IFR flight plan based on the fact that there were no advisories. After the encounter, the pilot could not understand why a precaution or advisory regarding that system was not provided. There were no advisories in effect because, at the time of the briefing, none were warranted. The pilot had the clues—moist, unstable air; thunderstorms already reported—but put complete trust in a forecast that included no precautions or advisories.

The preceding examples illustrate several decisions. One resulted in a canceled flight, another a routine flight—although the route was changed—and the last was almost fatal. My intent is not to brag about my skills or criticize another individual. I hoped to show the process, based on available information, a knowledge of weather, and limitations that led to the decisions.

All too often, briefers hear pilots flying aircraft without storm detection equipment say, "Thunderstorms, ah; well I'd better go IFR." Not for me, thanks. I want to be clear of clouds where I can see and avoid convective activity. Pilots who fly in conditions favorable for thunderstorms without storm detection equipment and the knowledge to use it sooner or later will end up—more likely upside down—in a thunderstorm cell.

A popular aviation saying goes, "Aviation in itself is not inherently dangerous. But to an even greater degree than the sea, it is terribly unforgiving of any carelessness, incapacity or neglect."

The existence of an advisory, or lack thereof, does not relieve the pilot from using good judgment and applying personal limitations. Like all pilots, I have had on occasion to park my turbo Cessna 150 and take one of American's Boeing 727s. These instances lend credence to the axiom: "When you have time to spare, go by air; more time yet, take a jet." When you don't have the equipment or qualifications to handle the weather, don't go! This doesn't mean every time we hear a weather advisory we cancel, but we do take a close look at all available information.

The following are some dos and don'ts of thunderstorm avoidance:

- Do avoid by at least 20 miles any thunderstorm identified as severe or giving an intense radar echo; this is especially true under the anvil of a large cumulonimbus.
- Do clear the top of a severe thunderstorm by at least 1000 feet for each 10 knots of windspeed at the cloud top.
- Do regard as severe any thunderstorm with tops 35,000 feet or higher.

- Don't land or take off in the face of an approaching thunderstorm.
- Don't attempt to fly under a thunderstorm, even if you can see through to the other side.
- Don't try to circumnavigate thunderstorms covering more than half of the area, even with storm detection equipment.
- Don't attempt to enter areas of embedded thunderstorms without storm detection equipment.

If thunderstorm penetration cannot be avoided, the following steps are recommended before entering the storm:

- Tighten seat belts and shoulder harnesses. Secure all loose objects.
- Plan a course through the storm in the minimum of time.
- To avoid the most critical icing, establish a penetration altitude below the freezing level or above $-15°C$.
- Turn on pitot heat and carburetor or inlet heat.
- Establish power setting for turbulence penetration airspeed.
- Turn up cockpit lights to highest intensity to lessen danger of temporary blindness from lightning.
- Disengage autopilot altitude and speed hold.
- Tilt airborne radar antenna up and down occasionally. Tilting may help detect a hail shaft or a growing thunderstorm cell.

The following are some dos and don'ts during thunderstorm penetration:

- Do keep your eyes on the instruments. Looking outside increases the danger of lightning blindness.
- Do maintain a constant attitude. Let the aircraft ride with the turbulence. Maneuvers to maintain altitude increase gust loading.
- Don't change power settings.
- Don't turn back once in the thunderstorm. A straight course through the storm most likely will get you out of the hazards most quickly. Turning increases gust loading.

Three final words regarding thunderstorms: avoid, avoid, avoid!

When it comes to thunderstorms, microbursts, and windshear: A pilot's proper application of many resources—training, experience, visual references, cockpit instruments, weather reports, and weather forecasts—make avoidance possible.

6
Air masses in conflict

IN PREVIOUS CHAPTERS, WE DISCUSSED HOW AIR OF DIFFERENT properties does not tend to mix. In the last chapter, we learned about air masses, their source regions, and how they are modified. Now we can move on to see what happens when air masses collide. Because of the dynamic nature of weather, our discussion begins with generalizations, with "classic" examples of weather systems. Then, we can move on to specific examples of weather that affect North America and strategies to learn to safely negotiate its hazards.

When an air mass moves from its source region, it comes in contact with other air masses. In the zone of contact between air masses, temperature and moisture (density) and wind may change rapidly over short distances. This zone of relatively rapid change is called a frontal zone—usually referred to as a front. Temperature is the most important density factor; therefore, fronts almost invariably separate air masses of different temperatures. Other factors also distinguish a front, such as a pressure trough, change in wind direction, moisture differences, and cloud and precipitation forms.

Norwegian meteorologist Vilhelm Bjerknes and his son Jakob developed the polar front theory at the beginning of the twentieth century. World War I had begun and it was popular to use the language of the conflict. Thus, weather was described using words like fronts, advances, and retreats. Weather does resemble a war between air masses.

Fronts fall into one of four categories:

- Cold front
- Warm front
- Occluded front
- Stationary front

The earth's atmosphere is a giant heat exchanger, moving cold air down from the arctic and warm air up from the tropics. Typically in the northern hemisphere the cold air pushes down from the northwest, lifts, and replaces the warm air. The boundary where this action takes place is called a cold front. However, to accomplish this heat exchange, at some point, warm tropical air must replace the colder air. This exchange typically takes place ahead of the cold front as the warm air, moving from the south, overrides and replaces the colder, retreating air; this boundary is known as a warm front. Cold fronts move faster than warm fronts; sometimes the cold front overtakes the warm front and an occlusion occurs. This front is called an occluded front. When frontal speed decreases to 5 knots or less, they are labeled stationary. Fronts produce vertical motion from the surface to about the middle troposphere—about 20,000 feet. This action, the polar front, is more or less continuous around the world at middle latitudes, as illustrated in Fig. 3-1 and Fig. 3-2. From the satellite picture in Fig. 3-2, it's easy to see that portions of the polar front are more active than others. An active weather system is approaching the Pacific coast, with benign weather over most of the western states.

Fronts have sloping boundaries, which are much shallower than those illustrated in this chapter. An average slope is between 1 in 50 and 1 in 300. That is, for every mile a front extends above the ground, the slope extends 50 to 300 miles downstream. Frontal boundaries lose their identity above 20,000 feet.

The width of a frontal zone depends on wind and temperature differences between the two air masses. The greater the temperature difference—other things being equal—the narrower the front; the stronger the wind component along and behind the front, the narrower the front. For example, a front moving 35 to 45 knots with a temperature contrast of 20°C or more may have a frontal boundary of 1 mile or less at the surface. A front with only a few degrees difference in temperature in a weak circulation may be as much as 50 miles wide. The movement and effects of wide fronts, with low temperature differences and weak circulation, are difficult to forecast.

Frontal intensity is based on frontal speed, which is determined by the temperature gradient in the cold sector, the region of colder air at a frontal zone. The movement of fronts is affected by many factors, such as temperature, moisture, stability, terrain, and upper-level systems. A front with waves indicates weak low pressure centers or portions of the front moving at different speeds. A front with waves should be watched. The weak low pressure areas can intensify and cause significant weather. The closer the front is to the jet stream, the steeper the slope and typically the stronger the front.

Fronts run the spectrum from a complete lack of weather, to benign clouds that can be conquered by the novice instrument pilot, to fronts that spawn severe thunderstorms that no pilot or aircraft can negotiate. Each front—for that matter any weather system—

must be evaluated separately, and then a flight decision must be made based on the latest weather reports and forecast and the pilot's and aircraft's capabilities and limitations.

Often the exact location, and sometimes even the presence, of fronts is a matter of judgment. Additionally, fronts do not necessarily reach the surface; they might be found within layers aloft. This is especially true in the western United States and the Appalachians, where mountain ranges break up fronts. Therefore, there might be differences between the charted position of fronts and their locations as described in forecasts or plotted on charts.

The location of surface fronts can be obtained from the surface analysis chart or weather depiction chart. The surface analysis chart also provides front type, intensity, and intensity trend. A narrative description of the location of fronts is provided in the synopsis portion of the area forecasts and transcribed weather broadcast (TWEB) route forecast synopsis. Additional information on the location and movement of fronts, when associated with severe weather, are contained in convective SIGMETs, severe weather watch bulletins, and the convective outlook.

THE LIFE CYCLE OF A FRONT

Frontal formation is called *frontogensis.* This process has several mechanisms. One is the interaction of a polar air mass with the polar easterlies and a tropical air mass with the subtropical westerlies. Between these air masses is a convergence zone. As previously mentioned, occasionally fronts develop when a new wave forms on the westward end of the cold front due to varying frontal speed or terrain.

As illustrated in Fig. 3-1, between the subtropical westerlies and polar easterlies lies the more or less permanent polar front. This area is where most of the midlatitude cyclones—low pressure areas—develop.

Refer to Fig. 6-1. Figure 6-1a shows a segment of the polar front that is stationary. (On black-and-white weather maps, a stationary front is shown as a line with alternating triangles and semicircles. On color weather maps, it is drawn as an alternating red and blue line.) When winds are parallel to the polar front, there is little, if any, weather along the front. This is illustrated in the satellite image depicting the portion of the polar front over the intermountain region of the western United States (see Fig. 3-2). Note the almost total lack of clouds.

This type of flow, however, sets up a cyclonic windshear. Under the right conditions, a kink or wave forms along the front. Winds are no longer parallel and frontogensis begins, as illustrated in Fig. 6-1b. (The gray areas in Fig. 6-1 represent typical areas of precipitation.) The region of lowest pressure is at the boundary of the two air masses—the front. In Fig. 3-2 frontogensis appears to be developing in the eastern Pacific between the cloud mass approaching the Pacific coast and that overrunning the high north of Hawaii.

As the cold air mass of the polar air begins to push southward under the warmer air of the tropical air mass, a cold front develops. (On black-and-white weather maps, a cold front is shown as a line with triangles pointing in the direction of frontal movement. On color weather maps it is drawn as a blue line.) This is illustrated in Fig. 3-2. The cold

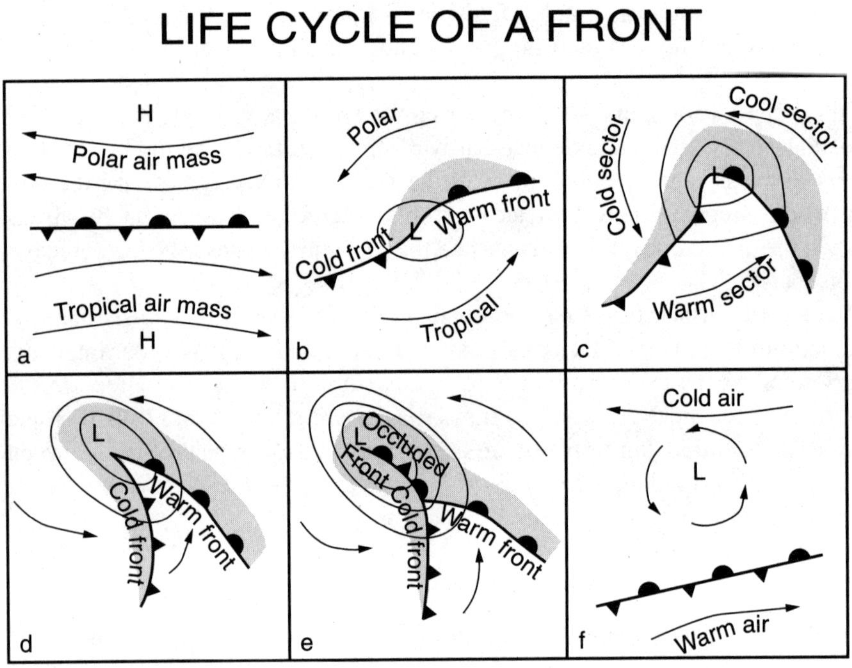

LIFE CYCLE OF A FRONT

Fig. 6-1. *Between the subtropical westerlies and polar easterlies lies the more or less permanent polar front.*

front is on a north-south line off the Washington, Oregon, and northern California coast. There is a distinct boundary between the cold and warm air, and the cloud band is relatively narrow.

As the cold air retreats on the north side of the low, the warm subtropical air mass moves in and replaces the colder air, as illustrated in Fig. 6-1c. This generates whirlpools between adjacent currents, and waves form along the polar front. As the advancing warm air mass overrides and replaces the retreating cool air, the boundary that forms is called a warm front. (On black-and-white weather maps a warm front is shown as a line with semicircles pointing in the direction of frontal movement. On color weather maps it is drawn as a red line.) That portion in Fig. 3-2 north of the polar front in southwestern Canada represents this area. Note the relatively large cloud band. The front is weak in Montana, with the cloud band some distance from the frontal boundary.

Note in Fig. 6-1c that we refer to the area of warm air as the warm sector and the cool air ahead of the warm front and cold air behind the cold front as the cool or cold sector.

Directed by the upper winds, the wave generally moves to the east or northeast. The central pressure lowers as cyclonic flow becomes stronger. Precipitation forms in a wide band ahead of the warm front and along a narrow band behind the cold front.

The faster-moving cold front constantly moves closer to the warm front. Eventually, the cold front overruns the warm front and the system becomes occluded—an occluded front. (On black-and-white weather maps an occluded front is shown as a line with alternating triangles and semicircles on the same side, indicating the direction of movement of the front. On color weather maps it is drawn as a purple line.) At this point, the storm is usually more intense with widespread clouds and precipitation, as illustrated in Fig. 6-1d and Fig. 6-1e. The area off the Canadian coast, westward to the low pressure center, in Fig. 3-2 represents the area of occlusion and the occluded front.

As temperatures and pressures equalize across a front, the front will dissipate, a process called *frontolysis.* Without the energy of the rising warm moist air, the old storm system dies out and gradually dissipates, as illustrated in Fig. 6-1f. (On black-and-white weather maps a stationary front is shown as a line with alternating semicircles and triangles on opposite sides. The semicircles point toward the cold sector, triangles toward the warm sector. On color weather maps a stationary front is drawn as an alternating blue and red line.)

The development or strengthening of a cyclone is called *cyclogenesis.* Over the United States, this development tends to occur over the Great Basin (the area between the Rockies and Sierra Nevada mountains, consisting of southeastern Oregon, southern Idaho, western Utah, and Nevada), the eastern slopes of the Rockies, the Gulf of Mexico, and the Atlantic Ocean. Some of these waves develop into huge storms, others dissipate within a day. The difference is the upper-level flow, which is discussed in the next chapter.

Weather associated with any front depends on both the slope of the front and the properties within each of the conflicting air masses. Vertical motion, usually upward, occurs above the frontal surface due to the air flowing along the slope. Thus, when a front forces conditionally unstable air upward, showers or thunderstorms form above the frontal zone. With abrupt lifting along a steeply sloping front, significant weather forms in a line near the surface position of the front. A more gradually sloping front may spread the unstable weather for many miles over the frontal surface. Thunderstorms may be embedded in an extensive area of stratified clouds. Moist, stable air overriding a frontal surface produces stratified clouds and steady precipitation.

COLD FRONTS

When a cold air mass replaces a warm air mass at or near the surface, it is by definition a *cold front.* Since cold air is more dense than warm air, the cold air flows under and replaces the warm air, forcing the warm air upward—a lifting mechanism. A cold front moves at about the speed of the component of the wind perpendicular to the front just above the friction layer between the surface and free air. (Recall the friction layer is from the surface to about 1000 feet over oceans and 2000 feet over land.) The area ahead of the cold front is called the warm sector and behind the front, the cold sector, as illustrated in Fig. 6-1.

Figure 6-2 shows a cold front with slightly unstable continental polar cold air underrunning warm, moist, unstable, maritime, tropical warm air. Abrupt lifting at the surface frontal position releases the instability into a line of thunderstorms. This area is where squall lines can develop. Note the relatively narrow area of weather associated with the cold front.

Towering cumulus may also develop ahead of the front due to surface heating in the warm air along with wave motion generated by the front. Recall our discussion of squall

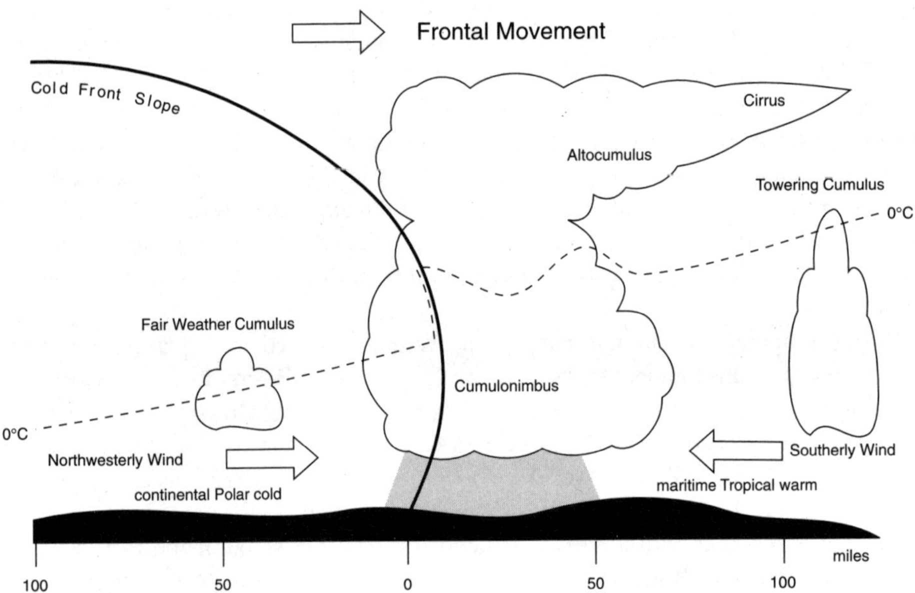

Fig. 6-2. *A cold front with slightly unstable cold air underrunning warm, moist, unstable air produces abrupt lifting and instability.*

lines from the last chapter. A towering cumulus cloud, technically *cumulus congestus,* is illustrated in Fig. 6-3. Note that the top of the cloud has not transformed into a smooth, fibrous, cirroform type. These clouds often develop into thunderstorms. Towering cumulus should be treated with the same respect as cumulonimbus.

The approach of the front is often signaled first by cirrus clouds that thicken into cirrostratus. Next appears a thickening band of altocumulus, similar to those shown in Fig. 4-4. These are quickly followed by cumulonimbus clouds, thunderstorms, heavy rain, and low ceilings and visibility.

Surface winds are typically out of the southern quadrant in the warm sector and out of the northwest quadrant in the cold sector. Precipitation is generally heavy and showery, associated with an unstable air mass. The freezing level lowers with the approach of the front, drops dramatically during frontal passage, and continues to lower in the cold air behind the front.

Active cold fronts are characterized by

- Fast movement
- Good visibility before and after frontal passage
- Turbulence in the frontal zone
- Relatively narrow band of icing in the frontal zone
- Sharp temperature changes over relatively small distances

- Relatively large changes in moisture content
- Large shifts in wind direction
- Strong pressure gradients
- Relatively small areas of heavy, showery precipitation.

Typically, the front passes rapidly with clearing skies and cool temperatures. Behind the front, fair-weather cumulus develop in the slightly unstable cold air. Fair weather cumulus are illustrated in Fig. 6-4. These clouds signify turbulence below and smooth air above. Pilots beware, even fair-weather cumulus above the freezing level can cause significant structural icing.

Figure 6-5 shows a cold front underrunning warm, moist, stable air. Stable stratified clouds form above the front. Note the stable cloud types: cirrostratus, altostratus, and nimbostratus. This succession of clouds also announces the approach of the front. The cold air is stable except where surface heating has created a very shallow convective layer, producing stratocumulus clouds behind the front.

Surface winds, again, are typically out of the southern quadrant in the warm sector and out of the northwest quadrant in the cold sector. Precipitation is generally light to moderate and steady, associated with a stable air mass. The freezing level lowers with the approach of the front, drops dramatically during frontal passage, and continues to lower in the cold air behind the front.

Weak or dry cold fronts are characterized by the lack of cloud cover and no precipitation. Both cold and warm air masses are dry and stable. The warm air must reach

Fig. 6-3. *Towering cumulus should be treated with the same respect as cumulonimbus.*

Fig. 6-4. *Fair-weather cumulus develop in the slightly unstable cold air behind the front.*

significant heights before condensation occurs. Clouds are typically a considerable distance from the surface position of the front and at high levels, the height depending on the moisture content of the warm air.

At the surface, cold frontal passage is characterized by a temperature decrease, a wind shift with often gusty winds, associated turbulence, and increasing pressure. With low-level moisture, weak cold fronts are characterized by stable clouds and low tops. In California on a flight from Van Nuys to San Luis Obispo, we were able to top the clouds of a weak, dissipating cold front at 6500 feet. With this front, cloud tops were actually lower than those produced by deep coastal stratus layers, which can at times exceed 8000 feet.

In the winter, the Gulf of Alaska region produces the most frequent cyclonic activity in the northern hemisphere. Low pressure is centered in the vicinity of the Aleutian Islands, and high pressure persists over the Great Basin of Nevada and Utah. The flow of air over the Pacific Northwest is primarily from the south and southwest. The favored storm track curves from the central Pacific northeastward towards Vancouver Island. These tracks frequently shift southward over Oregon and California and occasionally to southern California. Unlike the central and eastern United States, in the cold season the source of moisture is brought in with the cold air.

Fronts approaching from the northwest that penetrate southeastward are usually associated with colder, more unstable air masses and produce cumuliform clouds with showers and considerable clearing behind the fronts. These storms are accompanied by strong southerly winds ahead of the front, shifting to westerly with passage.

With winter storms, freezing levels typically range from several thousand feet in Washington and Oregon west of the Cascade Mountains to near the surface east of the Cascades. The temperature in the cold sector of the front may be higher than at the surface, which results in freezing rain or drizzle east of the Cascades and occasionally as far south as the interior mountains of Northern California. Freezing levels in California typically range from 5000 to 8000 feet in the north to 12,000 in the south.

Frontal systems south of the surface and upper-level low pressure centers crossing the Pacific generally weaken as they move on shore. Storm systems that approach from the west—the Central Pacific—tend to be weak. These systems often appear on the surface analysis chart as cold fronts with waves or stationary fronts. These systems might extend well into the Pacific. They bring poor, but seldom severe, weather with prolonged precipitation for several days. There is little chance for clearing between successive systems.

Weather systems approaching the Pacific coast from the southwest are likely to be associated with air masses of relatively higher temperature and moisture content—storms with the "pineapple connection" or the "pineapple express" (from the latitude of Hawaii). These storms are accompanied by substantial cloud layers and bands of precipitation. If the air is stable, widespread areas of steady, heavy precipitation occur; if the air is unstable, heavy rain, rainshowers, and thunderstorms result. Freezing levels are usually high,

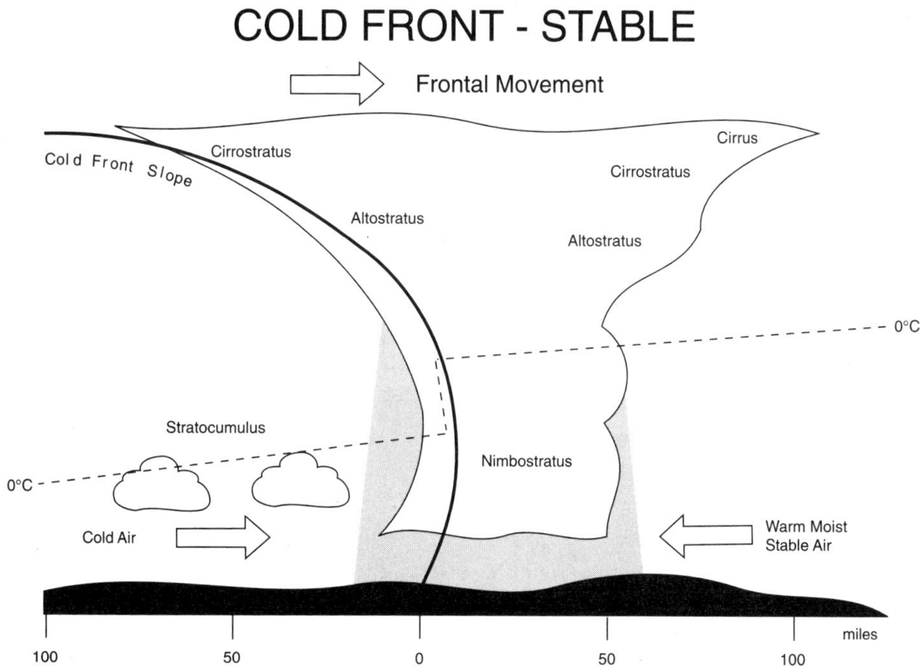

Fig. 6-5. *Cold fronts associated with stable air produce stratiform clouds over relatively short distances.*

often above 8000 to 12,000 feet, even into Northern California. They are often associated with, or only depicted, as an upper-level low. Poor weather can persist for days.

Pacific fronts enter the intermountain region—Idaho, Montana, Wyoming, Nevada, Utah, Colorado, Arizona and New Mexico—after a substantial overland journey. They tend to produce heavy precipitation in the form of rain or snow west of the mountains and light or no precipitation east of the ranges. Freezing levels are typically at or near the surface. These types of cold fronts frequently weaken over the southern portions of Utah and Nevada. They may produce little or no precipitation in these areas. They are often accompanied by gusty, shifting surface winds and blowing dust or sand.

Summertime fronts are usually weak, approaching the West Coast from the northwest. They frequently cause only drizzle along the northern coast and scattered showers or thunderstorms over the higher mountains. With their moisture dissipated in Washington and Oregon, they tend to be dry through the intermountain region. But their energy regenerates over the Rockies with the warm moist air from the Gulf of Mexico, which can produce severe weather in the Midwest and East.

Polar fronts, common in winter months, enter the intermountain region from the north. They are usually frigid and fast-moving. Under the influence of these fronts, blizzard conditions are not uncommon throughout the area. (A blizzard is a snow storm with winds in excess of 30 knots, visibility less than ¼ mile, and lasting three or more hours.)

Arctic fronts descending out of Canada, moving southward or westward, may result in cold arctic air pushing across the mountain barriers all the way to the coast. This normally only occurs a few times over a period of several years. Snow and freezing rain may accompany the front, or skies may be clear if the air masses contain little moisture.

Because of abundant low-level moisture and an upslope flow, the mountains normally take 24 to 36 hours longer than the coastal sections or valleys to clear following frontal passage. Surface frontal positions become difficult to locate as irregular mountain barriers are crossed. The frontal weather becomes diffuse as the air masses ahead of and behind the fronts are forced to ascend the windward slopes, intensifying the weather. On the leeward sides, descending air warms, and activity diminishes.

Between winter storms, strong high pressure often develops over the area, especially in the Great Basin. High pressure often results in the formation of fog in the valleys, Santa Ana winds in southern California, and the Chinook of the Rockies.

In the summer season, large changes occur in the circulatory pattern. The Aleutian low weakens and moves northward, as does the main storm track. The center of the Pacific high moves northwestward, and a ridge extends northeastward over the Pacific Northwest. The Great Basin high disappears, and a thermal low forms over the desert southwest with its trough extending northwestward towards the northern California coast and at times all the way to eastern Oregon and Washington.

Polar fronts still influence the weather, especially east of the Continental Divide. Here, moisture that drives the weather comes from a moist south to southeasterly flow from the Gulf of Mexico. This moisture can, at times, move all the way into the desert southwest.

East of the Rockies, cold fronts take on their more-classic depictions. Weather deteriorates with the approach of the front, then clears rapidly after frontal passage. Moisture that fuels the front comes from the maritime tropical air of the Gulf of Mexico and the Atlantic Ocean. Fronts are more severe in late spring and early summer because this is the season of greatest temperature difference, producing severe thunderstorms and tornadoes.

Arctic and polar air can, occasionally, reach the Gulf of Mexico or the Atlantic seaboard of the southeast United States. Freezing temperatures may reach southern Texas and Florida. As the cold air moves through Texas, it may lower temperatures by as much as 10° in a few hours. In this region the cold wind is known as a "Texas norther" or "blue norther," especially when accompanied by snow.

Occasionally a strong storm system develops near the province of Alberta in Canada. The storm moves rapidly east and southeast into the Great Lakes and then into the northeastern United States. Severe weather and blizzards accompany the system, colloquially referred to as the "Alberta clipper."

We have discussed how waves can form and storms intensify. One such storm is the "northeaster" or "nor'easter." They usually develop in the lower midlatitudes (30° to 40° north latitude) within 100 miles east or west of the Atlantic coast. An intense low pressure system progresses generally northward to northeastward along the East Coast of the United States and typically attains maximum intensity near New England and the Maritime Provinces of eastern Canada. These storms nearly always bring heavy precipitation in the form of rain or snow and frequently winds of 30 to 50 knots—occasionally 80 knots—out of the northeast; hence the name. They may occur at any time of year but are most frequent and violent between September and April.

Cold fronts are characterized by

- Primarily cumuliform clouds, but stratiform clouds are predominant in stable air
- Precipitation generally in narrow bands, showery, and heavy at times
- Icing above the freezing level
- Wind shifting from southwesterly to northwesterly with frontal passage
- Fronts moving southeasterly at about 20 to 25 knots
- Temperature decreasing with frontal passage
- Pressure falling as front approaches, then rising after frontal passage

WARM FRONTS

Cyclonic flow around the low in the cool sector causes the cool air to retreat. As the cool air retreats, warm air from the warm sector overrides and replaces the cool air. This boundary is called a warm front. Winds in the warm sector must exceed the speed of the warm front, which averages about 10 knots. The slope of a warm front is considerably less than that of a cold front because of density differences between the air masses. That is, typically the air masses along a warm front have less density difference than along a cold front boundary.

Figure 6-6 shows a warm front with warm, moist, unstable air overriding cool, stable air. Note the lifting distance along the shallow front is more gradual than along the cold front, as shown at the bottom of Fig. 6-4. Instability showers and thunderstorms are spread out above the frontal surface rather than in the line, as in Fig. 6-2. These

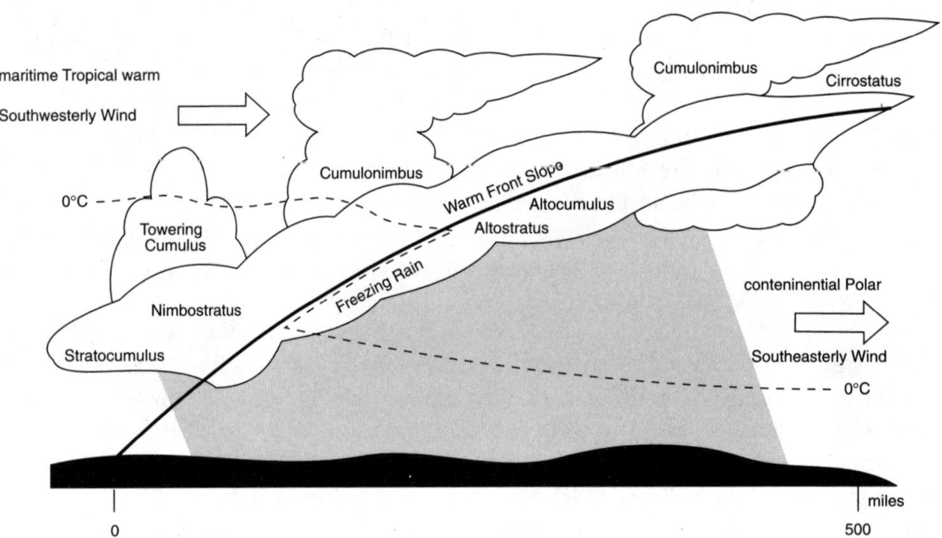

Fig. 6-6. *Instability showers and thunderstorms are spread out above the frontal surface rather than in a line, as with a cold front, and often are embedded.*

convective clouds may be embedded in a thick deck of stratiform clouds. Embedded thunderstorms are a very serious hazard to aircraft without storm detection equipment. Stratus fractus may form in the precipitation due to evaporation of water from the relatively warm rain and subsequent condensation in the cool air.

The approach of a warm front is often signaled by the appearance of cirrostratus, altocumulus, and altostratus clouds. These clouds lower to nimbostratus as the front nears. Tower cumulus and cumulonimbus are often hidden by the stratiform layers. Pilots flying on top are typically able to see the cumulus buildups.

Surface winds are typically out of the southern quadrant in the warm sector and out of the southeastern quadrant in the cool sector. Precipitation is generally light to moderate, steady, and widespread, with embedded areas of heavy, showery precipitation associated with the unstable air mass. The freezing level rises with the approach of the front and continues to rise along the frontal slope in an area of frontal inversion, as shown in Fig. 6-6. Frontal inversion can be a significant hazard associated with a warm front. Rain falling from above freezing temperatures forms large, supercooled water droplets in the freezing air below. This area of freezing rain can be the most significant icing hazard and affects both VFR and IFR flights.

After frontal passage aloft, the freezing level continues to rise in the warm air behind the front. Since the slope is shallow, pilots can expect to penetrate the frontal boundary some distance from the location of the front on the ground. With warm front passage, pressure and temperature rise, and skies begin to clear.

Figure 6-7 shows a warm front with warm, moist, stable air overriding cool, stable air. Here again, lifting along the shallow front is gradual. Cloud types are widespread and stratiform, associated with the stable air. The VFR pilot is faced with widespread areas of precipitation and low ceilings and visibility; the IFR pilot should fare much better, with icing the most significant hazard.

Surface winds are typically out of the southern quadrant in the warm sector and out of the southeastern quadrant in the cool sector. Precipitation is generally light to moderate and steady, associated with the stable air mass. The freezing level rises with the approach of the front and continues to rise along the slope of the front. Large, supercooled water droplets produce freezing rain or freezing drizzle. After frontal passage aloft, the freezing level continues to rise in the warm air behind the front.

Active warm fronts are characterized by

- Slow movement
- Poor visibility before frontal passage, usually good after frontal passage
- Smoothness, except in areas of embedded cumulonimbus
- Widespread areas of icing and freezing rain
- Large temperature changes over relatively large distances
- Relatively large changes in moisture content
- Moderate shifts in wind direction

WARM FRONT - STABLE

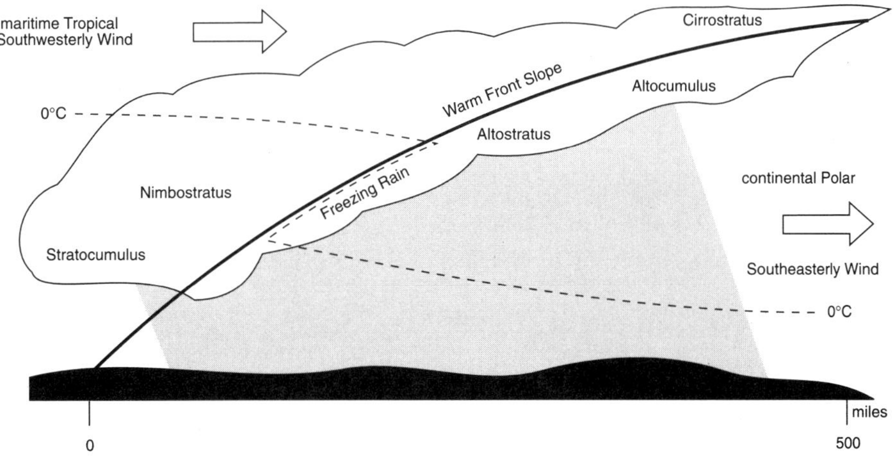

Fig. 6-7. *A warm front with warm, moist, stable air overriding cool, stable air produces stratiform clouds.*

- Moderate pressure gradients
- Relatively large areas of steady or showery, moderate to heavy precipitation

At the surface, the passage of a warm front is marked by an abrupt wind shift from southeast to southwest and an increase in pressure. This shift is usually associated with a clearing trend.

Weak warm fronts are characterized by the lack of cloud cover and no precipitation. If clouds form, they are high and a great distance from the front, as mentioned in our discussion of the life cycle of a front and illustrated in Fig. 3-2. High clouds with a dry, warm front clear several hours before frontal passage at the surface.

Warm fronts are relatively rare and usually weak in the western third of the United States, due to the fact that a moist, southerly flow is typically not found in this region. This does not mean, however, that warm fronts never exist in the region. I had taken my turbo Cessna 150 for its annual inspection from Lovelock to Winnemucca in Nevada. A warm front moving through the area produced thunderstorms and small hail. I had planned to fly back with the mail pilot in a twin Comanche. It wasn't to be. I had to come back by dog—a Greyhound to be exact (yes, I mean the bus)! Object lesson: My experienced airmail pilot knew a light twin without storm detection equipment had no business flying in an area of thunderstorms, possibly embedded.

East of the Continental Divide is where warm fronts come into their own. Here is where we can expect the classic warm fronts as illustrated in Fig. 6-6 and Fig. 6-7. The moist, southerly flow provides the fuel for activity along both the cold front and warm front. In this region a warm front produces extensive areas of low ceilings, visibility, freezing precipitation, and thunderstorms.

Warm fronts are characterized by

- Primarily stratiform clouds, but cumuliform clouds are embedded in unstable air
- Generally widespread precipitation, steady and light to moderate, except with cumuliform clouds, which produce showery, heavy precipitation
- Icing above the freezing level, with areas of freezing rain or drizzle, especially during the winter season
- Wind shifting from southeasterly to southwesterly with frontal passage
- Front moving northeasterly at about 10 knots
- Temperature increasing with frontal passage
- Pressure falling as the front approaches, then rising after surface frontal passage

OCCLUDED FRONTS

We have mentioned that the speed of a cold front is greater than a warm front. So what happens when a cold front catches up with a warm front? An occlusion or occluded front develops when a cold front overtakes a warm front. Because of density differences, one of the fronts is forced aloft. An occluded front may be either a warm front occlusion or a cold front occlusion. The difference depends on the density of the air ahead of the warm front and behind the cold front.

(Please note that a warm front occlusion does not in itself indicate unstable air, nor does a cold front occlusion indicate stable conditions. Warm front occlusion may be unstable or stable, just as cold front occlusion may be unstable or stable.)

Figure 6-8 shows a warm front occlusion. Here cool air is overriding cold air, forcing the cold front aloft. The cool air behind the cold front is less dense than the cold air ahead of the warm front; thus, the cold front is forced up the warm front slope. In this example, the cool air is stable, cold air stable, and warm air unstable. This type of occlusion normally forms in the western part of the North American continent. In the United States this occlusion occurs when a cold front, backed up by cool, maritime polar air from the North Pacific, moves over the continent and catches up with a warm front that is preceded by colder continental polar air. Cloudiness has features of both unstable cold and warm fronts. Maximum convective cloudiness is along the cold front aloft; stratified clouds with possible embedded thunderstorms develop above the warm front. Stratocumulus form in the stable cool air behind the frontal boundary. Stratus fractus form in the cold air with warm rain falling from above.

Surface winds are typically out of the northwest in the cool, cold front sector, southerly in the warm air mass above the warm front, and out of the southeastern quadrant in the cold air ahead of the warm front. Precipitation is generally light to moderate and steady, with embedded areas of heavy, showery precipitation associated with the unstable air mass. The freezing level is typical of a warm and cold front with unstable air.

Figure 6-9 shows a cold front occlusion. The cold air of the cold front replaces the cool air at the surface, forcing the warm front aloft. Here the dense cold air forces

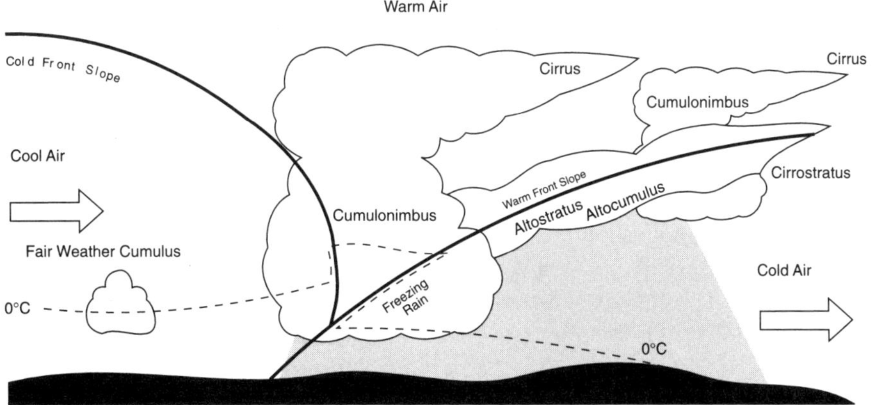

Fig. 6-8. *With a warm front occlusion, cool air overrides cold air, forcing the cold front aloft.*

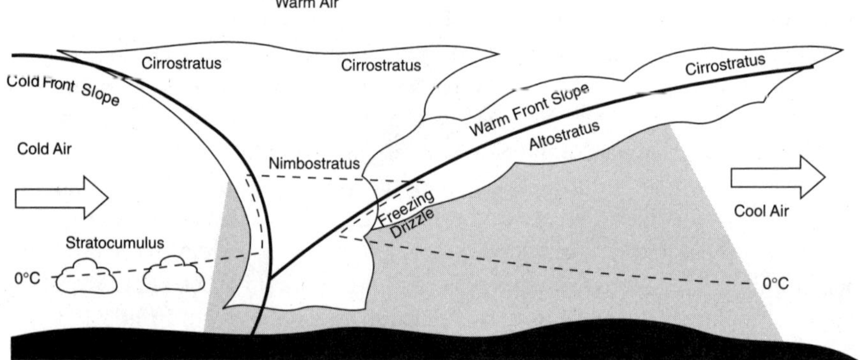

Fig. 6-9. *With a cold front occlusion, cold air from the cold front replaces the cool air at the surface, forcing the warm front aloft.*

the warm front slope up the cold front surface. Cold front occlusions are most common on the eastern two-thirds of the continent. In the eastern United States, the cold air on both sides comes from continental polar air. The cold air is moving south from its source region. The cool air ahead of the warm front has been over warmer terrain for a longer time; it has thus been modified and is relatively warmer than in the cold sector behind the cold front. The cold air is moderately stable, the cool air stable, and the warm air stable. Cloud types are stratiform associated with stable air and a combination of those contained in Fig. 6-5 and Fig. 6-7.

Surface winds and freezing level characteristics are similar to those of a cold front occlusion. Strong occluded fronts are characterized by weather associated with both strong cold and strong warm fronts. Like warm fronts, classic occluded fronts typically occur in the eastern two-thirds of the United States.

STATIONARY FRONTS

As the name implies, a stationary front has little or no movement—5 knots or less. Both cold and warm fronts can slow, showing little movement, and develop into stationary fronts. Slope is generally shallow, although it may be steep if density change across the front is large or wind distribution is favorable. Winds are parallel but opposite in direction, as illustrated in Fig. 6-1. Weather associated with a stationary front depends on its moisture, stability, and circulation. Clouds, if any, are stratiform, with little or no precipitation. When warm, moist air rides up and over the cold air, widespread stratiform clouds and light precipitation occurs. However, with moist, unstable air aloft, even the minimal low-level lifting associated with a stationary front can result in thunderstorms.

When warm, stable air from the west replaces cooler air to the east, the weather will be similar to that of a warm front with stable air. If the front begins to move, a warm front develops. Should colder air from the west replace warmer air to the east, weather will be similar to that of a cold front. If the air is unstable, cumulus clouds and thunderstorms may occur. Convective activity tends to be scattered and not as intense as that of an active cold front. Should the front begin to move, a cold front develops. As mentioned, waves in stationary fronts can develop, spawning new, intense weather systems.

Active stationary fronts are characterized by

- Little movement, less than 5 knots
- Poor visibility
- Little, if any, turbulence
- Weak temperature changes over relatively large distances
- Relatively small changes in moisture content
- Small shifts in wind direction
- Weak pressure gradients
- Relatively large areas of light, steady precipitation

Weak stationary fronts are characterized by the lack of cloud cover and no precipitation. Pilots flying through a weak stationary front may only experience frontal passage by an area of light turbulence and a small wind change.

Stationary fronts are characterized by

- Primarily stratiform clouds, but cumuliform clouds may develop with unstable warm air
- Generally widespread, steady, and light precipitation, if it occurs, except in cumuliform clouds.
- Freezing rain or drizzle during the winter season
- Generally light winds parallel to the front

STRATEGIES

When flying through a front, a pilot should be prepared for the changes in flight conditions from one air mass to the other. These changes are sometimes quite abrupt. Abrupt changes indicate a narrow frontal zone. At other times, the changes are very gradual, indicating a broad, weak, or diffuse frontal zone. Pilots should anticipate changes in cloud cover, temperature, moisture, wind, and pressure when penetrating a frontal zone.

Frontal systems produce every weather hazard so far discussed, including high density altitude—remember the thunderstorm heat burst? Therefore, we will not repeat the general implications and strategies of turbulence, icing, thunderstorms, and low ceilings and visibility, but instead relate strategies specific to frontal zones.

Essentially, pilots have two options when dealing with frontal systems: penetrate or avoid. If a pilot elects to penetrate, there are several additional options. These options are directly related to pilot and aircraft capabilities. A pilot may elect to fly over the front, penetrate at high or low altitude, or attempt to fly under the front. The pilot may want to fly through the front at a 90° angle to reduce exposure or accept the consequences and fly parallel to the frontal band.

Our route from Tulsa, Oklahoma, to Springfield, Illinois, was dominated by a weak cold front. Weather reports and forecasts indicated relatively clear conditions on both ends of the route and tops to about 6000 feet. Beneath the clouds, conditions were turbulent, with marginal ceilings and visibility. The decision was clear: With an instrument-rated and current pilot and an aircraft with instrument capability, VFR over the top was acceptable. We filed VFR and the flight was completed without incident.

Should flight on top be tried by a noninstrument-rated pilot? You have to evaluate the risk. An engine failure could lead to disaster. Navigational error or instrument malfunction could lead to getting caught on top. If this should occur, it cannot be overemphasized that the pilot should obtain assistance from ATC as soon as possible.

Depending on the type of front and associated weather, a pilot may elect to penetrate the front. A stable Pacific cold front was along our route from Van Nuys to San Francisco. We were flying a Cessna 172. Thunderstorms were neither reported nor forecast. The freezing level was forecast to be around 9000 feet in the south and 6000 feet in the north. We flight-planned along the coastal route, where minimum altitudes were 6000 feet or lower. Under such circumstances, flight over the coastal mountains into California's Central Valley, with minimum altitudes around 10,000 feet, was out of the question.

A pilot may wish to avoid frontal weather by either flying around the front or waiting for frontal passage. I have already related our decision on the flight from St. Louis to Kansas City with an active front producing thunderstorms. Flight decisions must be based on the capability of the pilot and aircraft, as well as the weather. In that case, frontal thunderstorms made the flight decision easy—no go!

There are really only two ways to avoid or reduce the exposure to frontal turbulence: Fly above the front or penetrate the front as close to perpendicular as possible. Penetrating a front at a 90° angle may work for a cold or stationary front but may not be satisfactory for a warm or occluded front because of the extensive areas of frontal weather associated with these fronts.

Icing problems exist mostly with winter fronts. With flight altitudes often governed by terrain clearance, freezing level heights are a real concern in flight planning. Additionally, in the mountains and high plateaus, the freezing level is often at the surface. In the summer, high freezing levels may create icing conditions into the lower flight levels.

Usually the solution to the icing problem is to fly high or low. Fly above the area of potential icing or below the freezing level. These options might not be possible in aircraft with marginal performance, especially in the western United States with its high minimum en route altitudes. Another option is to avoid visible precipitation; that is, remain clear of clouds and areas of precipitation when temperatures are less than 0°C.

Let's review a flight penetrating a cold front from the warm sector (A) to the cold sector (B), as illustrated in Fig. 6-10. Our pilot has decided to penetrate the front below

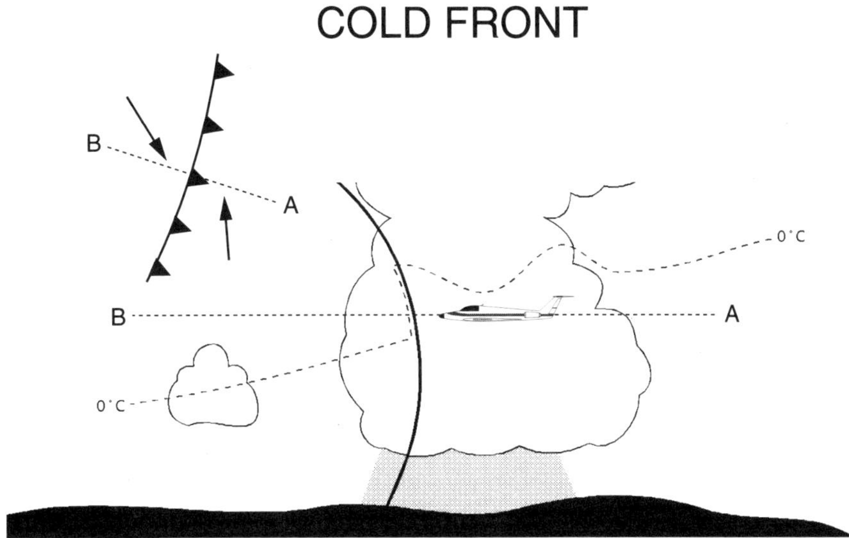

Fig. 6-10. *Frontal icing can be avoided by flying below the freezing level or avoiding areas of clouds and precipitation.*

the freezing level in the warm sector. The pilot could expect no icing problem until crossing the frontal boundary, as indicated by the 0° isotherm in Fig. 6-10. However, icing could be expected after penetrating the frontal surface aloft.

How would the pilot recognize frontal passage? Frontal passage aloft typically occurs at some point beyond the location of the front on the surface. It would be indicated by a change from a left crosswind to a right crosswind, as illustrated by the wind arrows in Fig. 6-10, and a decrease in outside air temperature. Pressure would decrease until reaching the frontal boundary; then it would increase. The pressure change may be significant, and the pilot would be well advised to obtain frequent altimeter setting updates. With a relatively steep frontal surface, the pilot could expect rapidly improving conditions after penetrating the frontal zone. Upon frontal slope penetration, the pilot would have the three icing options mentioned in Chapter 3: Climb to colder air above, descend to warmer air below, or reverse course to warmer air behind.

Essentially, the opposite scenario would be true for a pilot penetrating from the cold sector to the warm sector. Upon frontal slope penetration, somewhere prior to the location of the front on the surface, the pilot would experience a change from a left crosswind component to a right crosswind—the same as a flight from A to B. But the pilot would experience an increase in temperature and could expect a considerable amount of poor weather before exiting the frontal weather band.

With strong cold fronts, ceilings and visibility are generally good except in areas of heavy showers. Fog is unlikely because of gustiness and strong winds.

A VFR pilot might elect to fly under the front. The pilot could expect a relatively narrow band of sometimes heavy precipitation. The pilot could expect a relatively narrow area of low ceilings and visibility. With an active front, such as the one we encountered on our flight from St. Louis to Kansas City, the only option was to wait on the ground. However, with care, a weak front may be negotiated.

We had remained overnight in Huntington, West Virginia. Our next leg was to Cincinnati, Ohio, to repair the radio. A weak cold front was moving through the area. Without instrument flight capability, going through or over the front was not an option. Typical of a weak front, ceilings and visibility were poor, and because of the large frontal boundary, frontal passage was difficult to determine. We knew the weather in Cincinnati was clear. After departure we followed the river (IFR: I follow rivers) and had to negotiate about 10 miles of poor weather before getting into clear. It can be potentially dangerous flying at low altitudes in poor visibility with all the cables and catenaries crossing the river. Exiting the frontal boundary the weather was scattered stratocumulus with unrestricted visibility.

Now let's review a flight penetrating a warm front from the cool sector to the warm sector, as illustrated in Fig. 6-11. The pilot has elected to fly well above the freezing level, thus minimizing the risk of icing. However, entering the frontal inversion, the pilot would expect to encounter freezing rain or drizzle—a very hazardous icing zone. Here the pilot has a fourth option—climb to warmer air above. After penetrating the frontal surface, icing would no longer be a problem. The pilot could expect to penetrate the frontal slope well in advance of the surface front location. Frontal slope penetration would be indicated by a wind shift from southeast to southwest and an increase in outside air temperature. Like a cold front, pressure would decrease until reaching the frontal boundary, then rise. The pressure change would not be as significant as with a cold front,

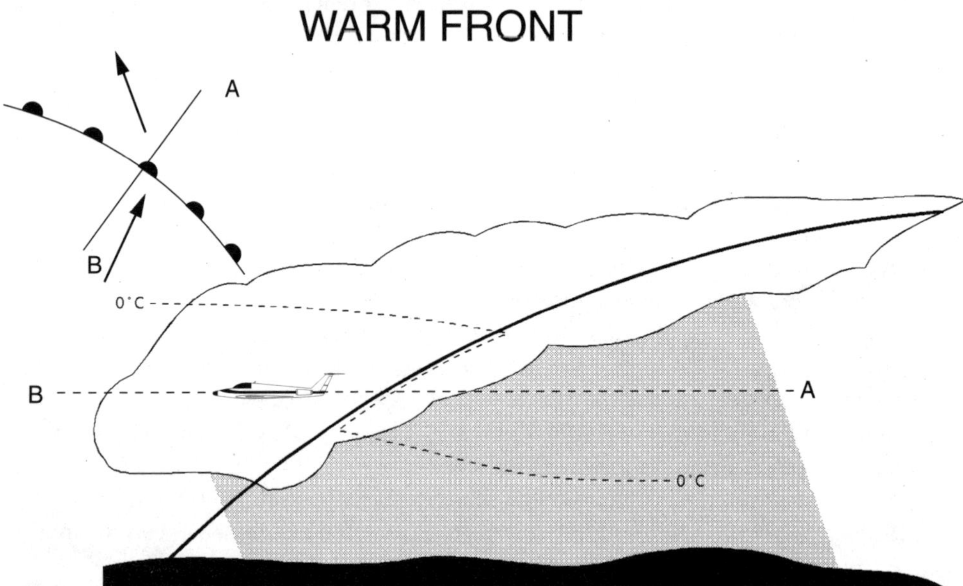

Fig. 6-11. *When penetrating a warm front, pilots must be prepared to contend with extensive areas of clouds and precipitation.*

but the pilot would still be well advised to obtain frequent altimeter settings updates. After frontal penetration, the pilot could expect extensive clouds and precipitation before exiting the frontal weather.

Again, the opposite scenario would be true for a pilot penetrating from the warm sector to the cool sector. In our example, the pilot would be well below the freezing level until encountering the frontal slope, located well ahead of the front's surface position. Frontal penetration would be indicated by an increased right crosswind component and a drop in temperature. The pilot may encounter freezing precipitation with an extensive area of clouds to negotiate before exiting the frontal weather.

With warm fronts, VFR pilots have to contend with widespread areas of low ceilings and visibility and the possibility of freezing precipitation. Pilots must use caution under these circumstances and have a suitable alternate airport close at hand during the entire flight.

As previously mentioned, weather and flight conditions in an occluded front will have characteristics of both cold and warm fronts.

Thunderstorms with all their ominous hazards are to be avoided. For aircraft with limited performance and without storm detection equipment, the only sane solution is to remain on the ground if these storms cannot be visually circumnavigated. In the last chapter we discussed these implications and provided a number of scenarios, strategies, and solutions.

With stationary fronts, flight is generally smooth, except when cumuliform clouds are present. Low clouds, ceilings, and visibility may persist in the cold sector.

The preflight briefing should provide the location and expected movement of frontal systems along a proposed route of flight. However, sometimes these buggers have minds of their own—that's the weather, not the forecasters.

By observing winds aloft and weather conditions, especially cloud types, we can often get a clue that the forecast isn't going according to plan. If we're IFR, our observations of wind direction and speed and outside air temperature can provide the same clues.

We should check with Flight Watch for updated weather reports and forecast. In this way, we can track the accuracy of the forecast and, especially at the first signs of unforecast conditions, consult Flight Watch for changing conditions.

Clues to unforecast frontal movement are clouds and wind and pressure changes either before or after they should have occurred. For example, if we're expecting a cold front to move beyond our destination, but current conditions are not improving as rapidly as forecast, the front has slowed and we had better find an alternate. The same would be true if we're attempting to beat a cold front to our destination. If conditions are deteriorating more rapidly than forecast, the front has accelerated—look out! IFR alternate minimums and fuel reserves take these factors into account. VFR pilots (with the exception of minimal fuel reserves) are pretty much on their own. It makes no sense to plunge into marginal or deteriorating conditions without adequate fuel reserves and suitable alternates. The excuse, "I didn't know conditions were that bad," is just that, an excuse!

We have our own eyes and the resources of ATC, FSS, and Flight Watch. The only reasons for getting caught in adverse weather are poor planning and failure to update weather en route.

7
Where's the front?

CONSIDERABLE MISUNDERSTANDING ARISES BECAUSE MANY aviation weather texts fail to adequately describe and explain nonfrontal weather-producing systems. A pilot presented with such a situation all too often asks the briefer, "Where's the front?" or, "When will the front pass?" only to be told, "There is no front." Recall that weather occurs at all altitudes within the troposphere and sometimes into the lower stratosphere; the surface analysis chart often cannot solely explain the weather, even that occurring at or near the surface. Upper-level troughs, as we shall see, play a key role in the evolution of weather and weather systems.

Upper-level weather systems tend to modify and direct surface weather. They can intensify or stabilize conditions at the surface, cause thunderstorms to occur, and enhance or retard the intensity of frontal zones. Upper-level weather systems can cause severe conditions at the surface or dampen or cancel out the vertical motion required to produce weather. The point that not all weather is caused by frontal systems is one theme of this chapter. In fact, nonfrontal weather-producing systems have considerable influence on surface conditions. The hurricane is the ultimate example.

We had remained overnight in Las Vegas because of the weather. The next day was also blustery. The route to Van Nuys was plagued with low clouds, mountain obscuration, and turbulence. I went into the Las Vegas Flight Service Station for a briefing. After providing the briefer with the necessary background information for the flight, this little old codger replied, "Well, you aren't going today!" This individual's technique was so obnoxious that without even thinking I replied, "Oh yes I am!" And I hadn't even looked at the weather.

As is often the case in the winter, an upper-level low was over the area. There were no associated fronts, and this kind of system tends to bring poor weather for days. The briefer was correct in part—we were not going to fly a direct course to Van Nuys. In addition to low ceilings and visibility, the route was dominated with scattered rain showers and isolated thunderstorms. However, a careful check of the weather showed that a route from Las Vegas to Needles, California, then to Daggatt and Palmdale was feasible. This route has the lowest terrain and plenty of alternate airfields. By circumnavigating the rain showers, we had a flight that was mostly smooth and without incident.

We had planned to attend the AOPA convention in San Antonio, Texas. From Van Nuys we made it as far as El Paso. The weather between El Paso and San Antonio was dominated by an upper-level low, producing low ceiling and visibility, icing, and embedded thunderstorms. Sometimes it's just not meant to be. We had to leave our turbo Cessna 150 and take one of American's 727s the rest of the way. Without ice protection and storm detection equipment, flight—VFR or IFR—under these conditions was out of the question.

The preceding incidents illustrate two upper-level weather systems. A careful check of the weather indicated one was a "go" and the other "no-go." Based on the weather, the aircraft's equipment, and a complete weather briefing, there really were no other rational decisions possible. When the weather is unflyable, don't go!

We have already mentioned that there are four to seven major upper waves in each hemisphere circling the globe. Embedded within this flow are shortwave troughs and ridges. In the summer, it's common to find longwave troughs near the coasts of the United States; in winter, one is commonly in the central portion of the country. Troughs allow the transfer of colder air down from the north and warmer air up from the south. The strength of a trough is determined by its windspeed, temperature differences across the trough, and speed of movement. With strong winds and large temperature differences, troughs produce more violent weather, such as squall lines and severe thunderstorms, typically in the spring and summer.

When a shortwave trough moves through a longwave trough, upward vertical motion is amplified, sometimes even producing upward vertical motion as it moves through a ridge. Shortwaves can be strong vertical motion producers. Most surface lows and frontal systems are associated with upper-level shortwave troughs. Because shortwave troughs move faster than associated surface frontal systems, a front can become stationary and left behind only to become active again as the next shortwave approaches. In addition to upper troughs and ridges, upper-level closed lows play an important part in the weather picture, along with upper-level highs.

Like surface low and high pressure areas, lows and highs develop aloft. These pressure systems have considerable effect on surface weather, as well as that aloft. Cut-off lows are powerful vertical motion producers. They can cause severe surface weather, as well as enhance the development of thunderstorms and produce severe clear air turbulence.

Highs aloft produce their own type of weather patterns. They can block weather systems, causing droughts and producing hellish-like heat waves, as in the classic case of the "Dust Bowl" of the 1930s.

As a general rule, locations north of a jet stream associated with a surface front are likely to be cold and stormy; locations south of this boundary tend to be warm and dry. Because of its influence on weather at the surface, as well as aloft, we revisit the jet stream in this chapter on more than one occasion.

Anything that spins has vorticity. In meteorology the term *vorticity* refers to the tendency of the air to spin. Vorticity is a powerful vertical motion producer. Like other atmospheric properties, it can be advected by the upper winds. Vorticity explains strengthening or weakening weather systems, the development of low pressure areas, and the existence of clear sky associated with weather systems such as upper-level lows and hurricanes.

Upper-level convergence and divergence occurs when air piles up or spreads out over a region. Convergence produces downward vertical motion, and divergence produces upward vertical motion. Upper-level convergence and divergence explain one reason for intensifying surface high pressure areas and deepening surface lows. Convergence and divergence are also associated with the jet stream. An area of upper-level divergence may be all that's needed to trigger thunderstorms and severe weather.

The hurricane is probably the best example of a nonfrontal weather-producing system. Hurricanes produce just about every kind of aviation weather hazard, covering thousands of square miles. Although these storms develop in the tropics, their devastation can reach well into the midlatitudes of the United States. With the exception of the Pacific Northwest coast, their weather can affect all coastal areas, and their remnants can reach hundreds of miles inland.

We discussed the development of frontal systems from a surface perspective in Chapter 6 and the effects of upper-level weather systems in this chapter. In the next section of this chapter, we put all of our knowledge of surface and upper-level weather systems together. The discussion contains a three-dimensional view of the weather systems.

Like previous chapters, we conclude with strategies to deal with upper-level weather systems. The discussion provides real-world solutions to the hazards of these phenomena.

UPPER LEVEL TROUGHS AND RIDGES

Forecasters love troughs. They use this term endlessly to explain all types of weather. They hide behind the term when a forecast goes sour, often blaming a busted forecast on a trough that is too strong, too weak, too fast, or too slow. Why do they love 'em so? Let's find out.

Global, or longwave, Rossby waves, with their associated ridges and troughs, extend for thousands of miles. A typical wave has a wavelength of from 50° to 120° of longitude. There is no temperature advection; isotherms—lines of equal temperature—are parallel to the contours. Longwaves generally move eastward at up to 15 knots but can remain stationary for days or even retreat. Air within the waves, however, moves at a much greater speed.

Figure 7-1 illustrates the flow through one of these longwaves. The figure is divided into two sections. The top portion shows the wave looking down from above. The lower section shows a profile, or horizontal, view of the air as it moves through the wave. The arrows show the direction of air flow.

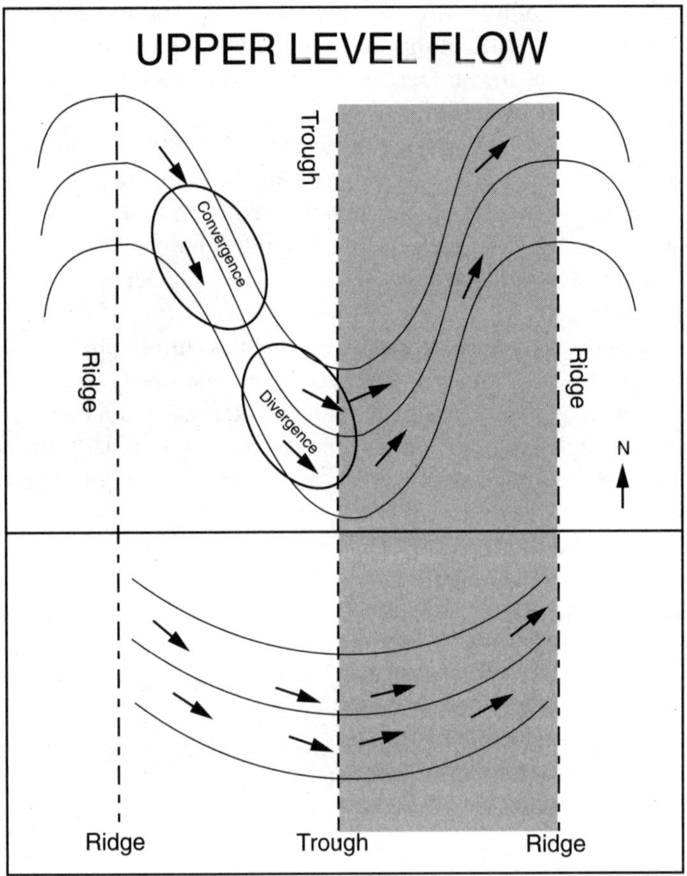

Fig. 7-1. *Upper-level waves produce downward vertical motion in the ridge-to-trough flow and upward vertical motion in the trough-to-ridge flow.*

In the northern hemisphere, flow through a longwave is generally from west to east, as illustrated in Fig. 7-1. The air flows parallel to the contours represented by the solid, curved lines in the top portion of the figure. As previously mentioned, air descends in the ridge-to-trough flow and rises in the trough-to-ridge flow. In other words, the longwave produces both downward and upward vertical motion. The gray, shaded area in Fig. 7-1 shows the area of upward vertical motion.

We can better visualize vertical flow within a wave by discussing the lower portion of Fig. 7-1. The solid, curved lines represent constant pressure surfaces. In the ridge-to-trough flow, you can easily see that the air is descending as it moves from the ridge to the trough. The flow changes to upward vertical motion in the trough-to-ridge portion of the wave.

Figure 7-2 is an infrared (IR) satellite picture showing a longwave. The ridge axis is on a north-south line through the center of the picture. The center of the upper-level high is over Baja California—note the absence of clouds in this area. The image shows extensive clouds in the trough-to-ridge flow. Note the clear skies over the central United States in the ridge-to-trough flow. This pattern often brings clear skies and strong, cold winds from Canada and the Arctic to the lower 48 states.

Moisture and clouds sometimes spill over the top or move through the ridge. When this occurs, weather types sometimes refer to it as a "dirty ridge." An example can be seen in Fig. 7-2 over the states of Montana and Wyoming. I'm not sure if "dirty ridge" refers to the unsettled character of the weather or the fact that forecasting these conditions is difficult. Several years ago we had just such a situation along the West Coast. Forecasters commented that rain fell from every cloud. A frustrated prognosticator wrote in an internal NWS product that they had finally determined the difference between a trough and a ridge. "Cold rain falls from a trough and warm rain from a ridge." This would seem to prove the proposition that weather forecasting is as much an art as a science.

Shortwaves, embedded in the overall flow, tend to pass through the longwave pattern at speeds of 20 to 40 knots. Shortwaves have wavelengths of from 1° to 40° of longitude. Shortwaves usually produce thermal advection—isotherms cross contours at some angle. Typically, cold-air advection occurs to the west of the trough axis and warm-air advection, to the east. Most surface lows and frontal systems are associated with upper-level shortwave troughs. Upper troughs, as we shall see, are a key to the evolution of weather systems.

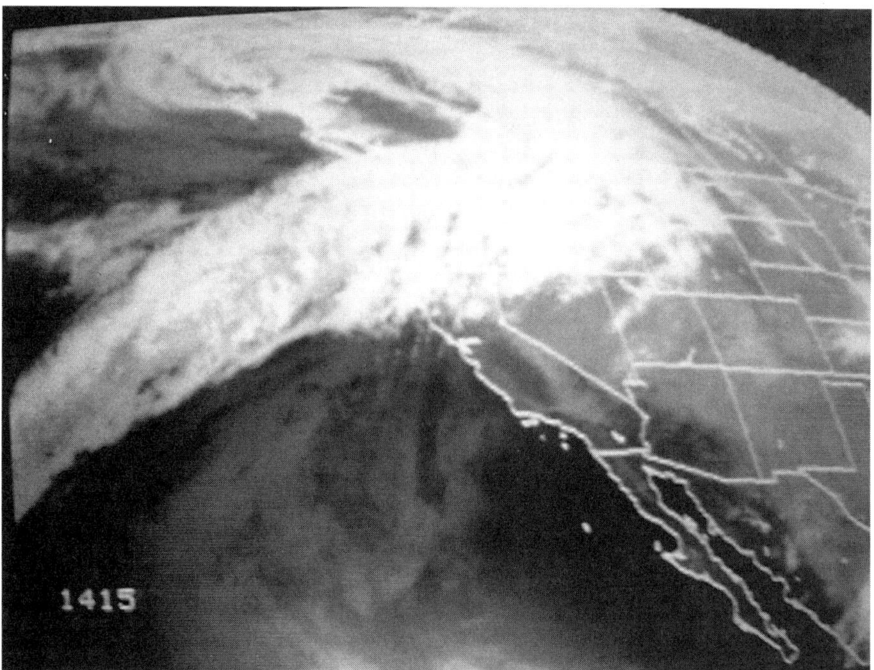

Fig. 7-2. *The IR image shows extensive clouds in the left half of the picture in the trough-to-ridge flow.*

How do shortwaves develop? Let's use the following analogy. Air, like water, behaves like a fluid. Let's think of the upper portion of Fig. 7-1 as a river. Water, like air, flows swiftly through the large meanders. Similar to a raging river, the flow comes against one bank and then the other and may be disrupted by debris in the river bed. To atmospheric flow, major mountain ranges, like debris in the river bed, affect the flow, causing eddies to develop in the flow that move downstream at about the same speed as the overall flow. These eddies correspond to shortwaves in the atmosphere. (Some have recently proposed that the ice age was, at least in part, due to a change in the jet stream caused by the development of the Himalayan Mountains.)

Refer to Fig. 7-3. Again the upper portion of the illustration is a vertical view, the lower portion a horizontal view, and the gray, shaded area represents upward vertical mo-

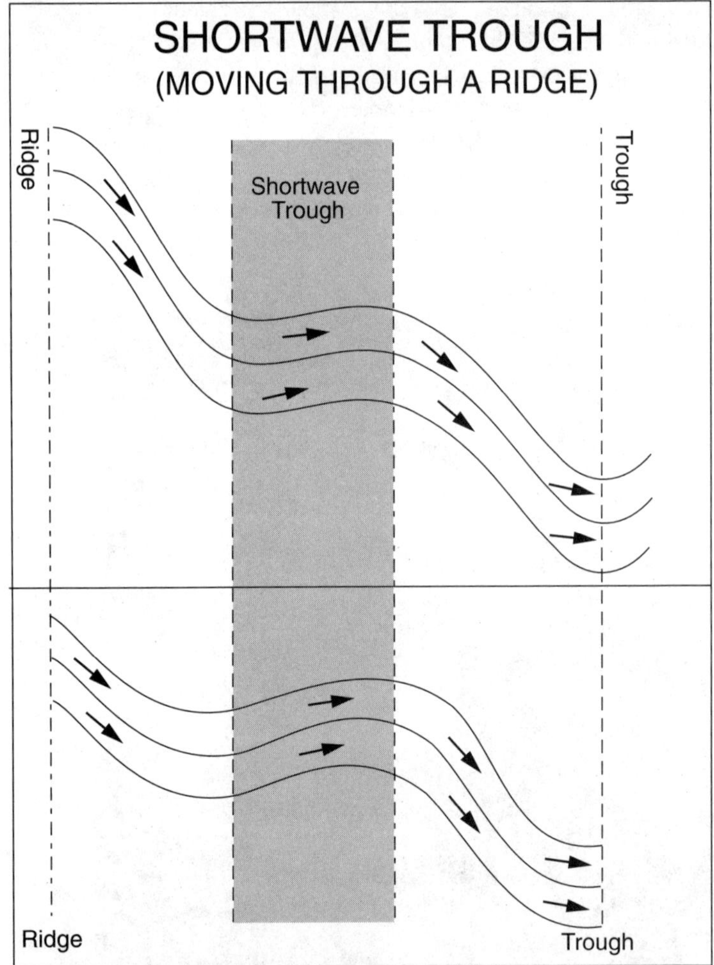

Fig. 7-3. *As a shortwave trough moves through a longwave ridge-to-trough flow, an area of upward vertical motion can be created.*

tion. As a shortwave trough moves through a longwave ridge-to-trough flow, an area of upward vertical motion can be created. Notice in the lower portion of Fig. 7-3 that in the shortwave trough to shortwave ridge flow, there is upward vertical movement. Thus, as the shortwave trough moves through a ridge, clouds and precipitation can occur. The extent of clouds and precipitation, if any, depend on the relative strengths of the shortwave and longwave.

Conversely, as a shortwave trough moves through a longwave trough-to-ridge flow, upward vertical motion is enhanced (Fig. 7-4). Typically, a shortwave ridge follows the shortwave trough. The end result, as illustrated in Fig. 7-4, is enhanced vertical motion in the gray, shaded areas of the illustration and even an area of reduced vertical motion in the shortwave ridge-to-trough flow. Again, the amount of clouds and precipitation depend on the relative strengths of the troughs.

The location of upper-level troughs and ridges can be found on constant pressure charts.

Upper-level troughs and ridges transport, or advect, atmospheric properties aloft. Cold-air advection destabilizes the atmosphere at the 500-mb level. Conversely, warm-air advection at this level stabilizes the atmosphere. This is opposite to the effects of cold- and warm-air advection near the surface. Why? As cold air aloft is advected into an area, any rising air will be warmer than surrounding air. Therefore, cold-air advection aloft enhances the development of weather-like thunderstorms by promoting vertical development. Warm-air advection at this level stabilizes the atmosphere, strengthens high pressure ridges, and diminishes low pressure troughs.

How does warm- or cold-air advection occur? Recall that isotherms are lines of equal temperature. For warm- or cold-air advection to occur, wind must blow across isotherms. This mechanism allows for the transportation of cold or warm air. Isotherms and areas of cold- or warm-air advection are found on constant pressure charts.

With the approach of an upper-level trough, a period of 8 to 12 hours of poor weather can be expected. The surface front precedes the trough, usually bringing IFR weather. However, without a front, the upper trough or low might only bring marginal VFR conditions with localized areas of IFR. Under these conditions, VFR flight might be possible, except in mountainous areas that remain obscured in clouds and precipitation.

On the other hand, the absence of an upper-level trough tends to weaken and slow a front's progress. A ridge aloft even with a surface front does not tend to produce thunderstorms or severe weather because the ridge prevents the vertical development required. Figure 7-5 shows a weak surface front moving into Idaho and Nevada. Notice the almost complete absence of clouds in California, even though these fronts often appear on the surface analysis chart. This front has minimal upper-air support. A temperature difference marks the boundary, with insufficient moisture and lifting to produce clouds. It is not all that unusual to have cloud tops below 10,000 feet with weak fronts.

Two general rules are applied to the 500-mb chart. At latitudes of the United States, weather systems tend to produce precipitation north of the 564 contour line and snow

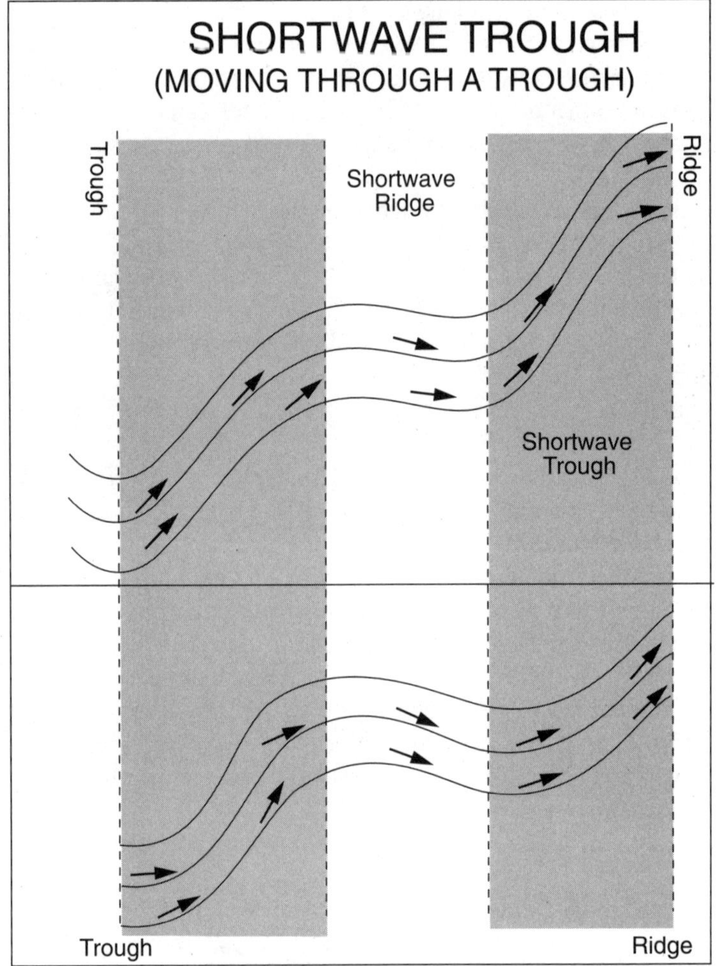

Fig. 7-4. *As a shortwave trough moves through a longwave trough-to-ridge flow, upward vertical motion is enhanced.*

north of the 540 contour line. These contour lines are sometime referred to as the "rain" and "snow" lines.

UPPER-LEVEL LOWS AND HIGHS

Like low and high pressure areas at the surface, upper-level lows and highs also occur. Their definition is the same. An upper-level low is an area completely surrounded by higher pressure and an upper-level high is an area surrounded by lower pressure. Recall that unlike the surface chart, upper-level charts depict weather patterns through the height of constant pressure surfaces. However, contours can be analyzed the same as isobars on the surface analysis chart.

At and above the 500-mb level, when an area of low pressure is completely surrounded by a contour, it is called a closed low or cut-off low. These cut-off lows, sometimes referred to as "cold lows aloft" or "upper-level lows," are an important winter weather feature. Under these conditions, low pressure is reflected from the surface to the tropopause. When the area of low pressure is vertical through the atmosphere, these storms tend to be powerful and erratic. Storms resulting from cut-off lows produce precipitation and low ceilings and visibility over widespread areas. Forecasting the formation of cut-off lows and their movement is difficult.

How do cut-off lows form? Typically an intense high pressure ridge is present over the eastern Pacific, with strong northerly or northwesterly winds aloft along the coast of the Pacific Northwest, with its associated jet stream. The jet typically works around to the south side of the trough and finally to the east side. After formation, the lows tend to move southward for the first 12 to 24 hours, in response to the strong northerly jet on the west side of the low. Cut-off lows tend to be slow-moving and erratic. Under their influence, weather can remain poor for several days or more. When the low develops over the Great Basin of Nevada and Utah, it is often called an "Ely Low." When these lows move over the north-central Rockies and plains, widespread blizzard conditions and heavy snow result. During summer months, closed lows aloft support the development of thunderstorms once surface lifting begins.

Fig. 7-5. *A weak surface front is moving into Idaho and Nevada; an upper-level ridge accounts for the almost complete absence of clouds in California.*

The following area forecast synopsis illustrates the impact of upper-level weather systems:

STRONG UPPER LOW OVER SOUTHWESTERN NEW MEXICO WILL MOVE TO NORTH CENTRAL TEXAS BY 22Z.

The 0900Z surface analysis chart showed weak surface high pressure over the western United States. However, the 1000Z weather depiction chart disclosed extensive areas of IFR and marginal VFR, with rain and snow occurring throughout New Mexico, Texas, and Oklahoma. The weather closed airports for days and was blamed for the deaths of dozens of people. The 500-mb analysis reveals the culprit. A deep upper-level low along the southern Arizona–New Mexico border and associated downstream trough produced devastating surface conditions.

During one episode an upper low drifted in the vicinity of Red Bluff, California, for five days. Pilots would call day after day wanting to know when the weather would clear. After a while, the common response became, "The low is forecast to move east out of the area tomorrow, but that's what they said yesterday."

Clear air turbulence (CAT) is common in the winter season. The jet stream, lows aloft, and sharp troughs can cause severe CAT. Sharp troughs can produce severe windshear turbulence as low as 8000 to 10,000 feet. CAT is implied by the jet stream, lows aloft, and sharp troughs, which can be found on constant pressure charts. Forecasts for CAT are contained in the SIGMETs and the AIRMET bulletin.

Upper-level troughs and lows tend to form bands of weather, as can be seen in Fig. 7-6. The upper-level low is centered off the central California coast. The front is moving into southern California and northern Baja, Mexico. Note the banks of clouds behind the front associated with the low. The weather deteriorates as a band moves through, then improves, only to deteriorate with the next band. The area forecast cannot, and does not, take this into account. Under these conditions, a pilot must be careful not to get suckered by a temporary improvement.

Upper-level highs typically bring warm, dry weather. They tend to block approaching weather systems and are sometime known as "blocking highs." Heat waves can develop when the subsidence from a upper-level high lingers over an area. A specific pattern associated with upper-level highs is called an "omega block." Upper-level flow resembles the Greek letter omega (Ω), with an upper-level high in the center of the pattern. An omega block is a strong blocking high that can remain stationary for days or weeks. Shortwaves and their associated weather ride up over the high. When this occurs, a drought can result. If the pattern continues for more than one season, severe droughts, such as the Dust Bowl of the 1930s, develop, although the effects of the Dust Bowl were aggravated by other factors. Climatologically, major droughts occur in the United States roughly every 22 years, mainly in the midwestern states. However, droughts do occur in other parts of the country.

Flying weather in areas dominated by upper-level highs is similar to conditions under the influence of a stable air mass—typically poor visibility. High temperatures result in high density altitude. Hot, dry conditions are ideal for brush and for-

Fig. 7-6. *Upper-level troughs and lows tend to form bands of weather.*

est fires with their resulting smoke layers. If the wind picks up, pilots often have to contend with widespread areas of blowing dust and sand.

At times upper-level highs develop over surface low pressure areas. Thermal lows and hurricanes exhibit warm-core lows, resulting in a relatively shallow low. Convergence occurs below the 500-mb level, due to thermal heating in the case of a thermal low and the release of latent heat in the case of a hurricane. Above the 500-mb level, divergence occurs. A net outflow occurs at the 500-mb level.

Lifting mechanisms have a cumulative effect. Upper-level troughs parallel to and behind a front intensify the storm. These fronts tend to be fast-moving. Figure 7-6 shows a strong cold front moving through Southern California. The satellite picture reveals a trough and low off the central California coast. These lows occur when cold air at the base of the trough is cut off from the cold air to the north. This closed circulation can lead to a circular jet stream. The weather in central California remains moist and unstable even though the surface front has passed. These systems tend to bring extended periods of weather, sometimes severe, in the form of bands one after the other.

Figure 7-7 illustrates distribution of constant pressure surfaces when high pressure aloft forms over low pressure near the surface and vice versa. This situation occurs with temperature advection, either at the surface or aloft. Essentially, when a low aloft forms over high surface pressure, there is relatively warm air near the surface and cold air aloft. When a high aloft forms over low surface pressure there is relatively cold air

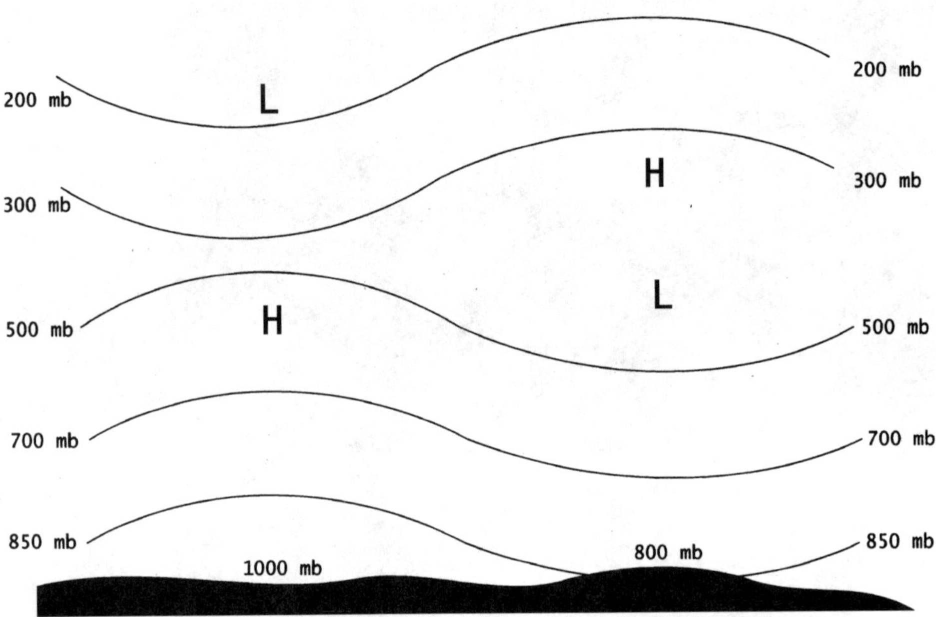

Fig. 7-7. *Temperature advection produces low pressure aloft over high pressure near the surface, and vice versa.*

near the surface and warm air aloft. Upper-level lows and highs can be found on constant pressure charts.

THE JET STREAM REVISITED

We included a section on the jet stream in Chapter 1. Because of its influence, this player has also been mentioned in sections on turbulence and thunderstorms. And, we talk about the jet stream again in our discussion of convergence and divergence. But, now we expand on its influence, primarily upon surface weather systems. We'll see just how this significant weather phenomena affects the overall weather picture.

The reason for the existence of the jet stream is the temperature contrast between the higher latitudes and the tropics. Meteorologists use complex computer models to predict the formation, location, and movement of the jet. Due to the complexity of the atmosphere, these predictions do not always come true. However, errors of timing are more prevalent than errors of occurrence. That is to say, the forecast almost always comes true, but the specific time of occurrence may be off.

Recall that the jet stream is not a continuous band of high winds. The jet stream is comprised of segments, sometimes called jet streaks or jetlets. The polar jet marks the boundary between polar air and warmer midlatitude air. The subtropical jet separates warm midlatitude air from warmer tropical air. At times there may be a third jet stream. The arctic jet, north of the polar jet, occurs between the boundary of extremely cold arctic air and cold polar air.

Small, sudden changes in the jet stream often coincide with the development of surface storm systems. These changes also affect the upper wind pattern both upstream and downstream from the occurrence. This has the effect of strengthening or weakening surface weather systems; we'll see how in our discussion of jet stream convergence and divergence.

Recall that, as a general rule, locations north of a jet stream associated with a surface front are likely to be cold and stormy; locations south of this boundary tend to be warm and dry. A jet embedded in a longwave can remain relatively stationary for weeks; this usually brings long periods of bad weather to the north of its location and good weather to the south. The movement of surface high and low pressure areas and fronts is related to the movement of the jet stream. Low pressure areas tend to move with the jet stream flow. As the wave with the jet passes, a ridge builds aloft, usually bringing high pressure and good weather. However, as high pressure builds, surface pressure gradients are often steep, causing strong, sometimes destructive, surface winds.

The jet stream typically has the following effect on surface weather systems. When the jet stream is parallel to a surface front, the front tends to be slow-moving and relatively weak. However, if the jet stream is perpendicular to a surface front, the front is apt to be fast-moving and strong.

Weather types often have interesting and obtuse terms to describe weather systems. For example, when a jet stream behind a trough increases the trough's strength, it is referred to as "digging." Therefore, a trough "digging" southward means a trough increasing in strength as it moves.

The location of the jet stream can most easily be found on the 300-mb and 200-mb constant pressure charts. Typically, the polar jet appears on the 300-mb chart and the subtropical jet on the 200-mb chart, because the polar jet is usually at a lower altitude; hence it appears on the lower-altitude pressure level chart. Which chart best reflects the jet varies with season and latitude. In the winter, look for the jet on the lower-altitude chart (300-mb), in the summer on the higher-altitude chart (200-mb). The location of the jet may also be revealed on satellite imagery by the presence of jet stream cirrus, jet streaks, and transverse lines or cloud trails. The forecast location, speed, and altitude of the jet stream appear on the high-altitude significant weather chart.

VORTICITY

Any nonmeteorologist pilot who wishes to better understand atmospheric phenomena requires a basic knowledge of vorticity. Although some pilot-meteorologists feel this subject is far too technical for pilots, I disagree. At the very least, pilots, especially those using direct user access terminals (DUATs), will come across this term in synopses, convective SIGMETs, and the convective outlook. An understanding of vorticity helps relate the fact that not all weather occurrences can be attributed to pressure and frontal systems alone, as displayed on weather charts.

Anything that spins has vorticity, which includes the earth. Vorticity is a mathematical term that refers to the tendency of the air to spin; the faster air spins, the greater its vorticity. Refer to Fig. 7-8. The spin can occur in any direction, but our concern is with air spinning vertically within the overall horizontal flow. A parcel of air that spins counterclockwise—cyclonically—has positive vorticity; a parcel that spins clockwise—anticyclonically—has negative vorticity.

How does vorticity occur? Divergence aloft must be compensated by convergence near the surface. For total mass of air to remain constant, the column must elongate vertically. So, what do we have? Upward vertical motion, independent of other vertical motion mechanisms. As the column stretches, spin—vorticity—increases, becoming more positive. Conversely, convergence aloft produces low-level divergence. The column flattens, its spin decreases, and its vorticity decreases—becoming less positive. The result: downward vertical motion.

The earth's vorticity is always positive in the northern hemisphere because the earth spins counterclockwise about its axis. An observer standing on the North Pole has maximum vertical spin, one revolution per day. An observer's vertical spin decreases as the observer moves toward the equator, becoming zero at the equator, like Coriolis force, which is maximum at the poles and zero at the equator. Vorticity is directly related to Coriolis force. The rate, or value, of the vorticity produced by the earth's rotation is, not surprisingly, known as the earth's vorticity.

Now consider the atmosphere, which is almost always in motion and generally has its own vorticity relative to the earth, or *relative vorticity.* The sum of the earth's vortic-

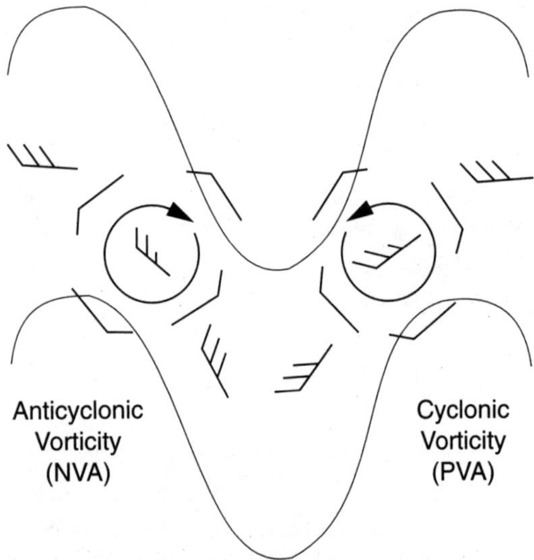

Fig. 7-8. *Areas of vorticity move through the overall flow.*

ity plus relative vorticity equals *absolute vorticity.* The value of absolute vorticity at midlatitudes almost always remains positive because of the earth's rotation.

Air moving through a ridge, spinning clockwise, gains anticyclonic relative vorticity. Air moving through a trough, spinning counterclockwise, gains cyclonic relative vorticity. Therefore, there tends to be downward vertical motion in ridge-to-trough flow and upward vertical motion in trough-to-ridge flow.

Vorticity is advected like other atmospheric properties. As illustrated in Fig. 7-8, areas of vorticity flow with the wind through the overall pattern. Their vorticity values remain unchanged. Therefore, these regions are known as areas of positive or negative vorticity advection. Pilots can expect to hear the terms *positive vorticity advection* (PVA) and *negative vorticity advection* (NVA) in reference to relative vorticity.

Positive vorticity advection indicates

- A trough or low moving into an area
- A ridge or high moving out of an area
- Upward vertical motion probably occurring
- Increasing cloud cover and precipitation

Negative vorticity advection indicates

- A ridge or high moving into an area
- A low or trough moving out of an area
- Downward vertical motion probably occurring
- Decreasing cloud cover

As regions of PVA sweep along within the overall flow, they represent microsystems that can rotate around synoptic-scale highs and lows. These areas may be referred to as "A VORT MAX" or "VORT LOBE." If an area of PVA moves over a stationary surface front, a wave can form and a storm can develop. An area of PVA might be all that's required (a lifting mechanism) to trigger thunderstorms when moisture and instability are available. On the other hand, NVA might retard or prevent thunderstorm development.

Vorticity explains the development of low pressure areas east of mountain barriers, particularly in eastern Colorado. These lows are most predominant east of the Rockies because of the high mountain elevations, sometimes referred to as "lee-side lows."

How do lee-side lows form? Across the Rockies there is typically a westerly flow. Air forced up the west slopes of the mountains is trapped below the tropopause. To compensate for the column's resulting compression, there is a decrease in absolute vorticity, which imparts on the flow an anticyclonic track to the southeast. As the air moves downslope, the column of air expands, and absolute vorticity increases. The air now begins a cyclonic track to the northeast, creating a low pressure area on the lee side of the mountain barrier.

Refer back to Fig. 7-6. We have mentioned that closed upper lows produce bands of weather. Vorticity explains this phenomena. In the areas of weather, there is PVA; in the relatively clear areas between the bands, NVA occurs. These areas of NVA are sometimes referred to as the "dry slot." Two distinctive dry slots can be seen in Fig. 7-6. One is located behind the cold front, the other within the closed low itself.

Absolute vorticity is analyzed at the 500-mb level. However, vorticity is not directly reflected on constant pressure charts. Even when the air is rotating within the overall system, balloon observation can only measure the mean, or average, windspeed. Refer to Fig. 7-8. Note that the windspeed before and after the areas of NVA and PVA are 30 knots. (The wind arrow shows the direction of the wind; wind barbs indicate speed. Each barb represents 10 knots and half barbs represent 5 knots.) There is an anticyclonic flow embedded in the ridge-to-trough flow of 10 knots. An area of embedded cyclonic flow of 10 knots has occurred in the trough-to-ridge flow. A weather balloon, or aircraft, flying through the area would only see the overall wind component of 25 knots (represented by the two full barbs and half barb on the wind arrows). The NWS produces composite height/vorticity charts. This chart series, with forecasts to 60 hours, depict areas and magnitude of both PVA and NVA. Although normally not directly available to pilots, 500-mb composite heights/vorticity charts are routinely transmitted. High values of absolute vorticity (greater than 16) have strong cyclonic rotation, indicating strong upward vertical motion, and low values (less than 6) have anticyclonic rotation, indicating strong downward vertical motion.

CONVERGENCE AND DIVERGENCE

Recall that surface convergence and divergence occur between the surface and about 10,000 feet. At the 500-mb level, halfway through the atmosphere, is a level of nondivergence. The air does not tend to converge or diverge at this level. Upper-level convergence and divergence occur above the 500-mb level.

Convergence in the upper atmosphere is the piling up of air over a region. Since air cannot be created or destroyed, it has to go somewhere. The only place for it to go is down. Therefore, convergence aloft is a downward vertical motion producer. Divergence is the spreading out of air over a region, which leaves a void or vacuum. Air is sucked up from below. Therefore, divergence aloft is an upward vertical motion producer.

Upper-level convergence and divergence can be seen on the 300-mb and 200-mb constant pressure charts.

Recall that at the surface, winds flow out of high pressure areas. In the upper atmosphere, the wind converges over a surface high. If upper-level convergence exceeds surface divergence, the high pressure will build and surface pressure increases. Conversely, at the surface, winds flow into low pressure areas. Above the low, in the upper atmosphere, the winds diverge. If upper-level divergence exceeds surface convergence, the storm system intensifies and surface pressure decreases.

What causes upper-level convergence and divergence? These phenomena may result from changes in wind direction and speed. Areas of convergence, also called *confluence,* occur in areas where the contours move closer together. The result is similar to the funneling effect of terrain. Windspeed increases. Areas of divergence, also called *difluence,* occur in areas where the contours move farther apart. In these areas, windspeed decreases. Areas of convergence and divergence are illustrated in Fig. 7-1.

Areas of convergence and divergence are also associated with the jet stream. Refer to Fig. 7-9. As air enters the jet maximum, it increases rapidly in speed over a relatively short distance due to an increase in pressure gradient force. The increase in pressure gradient temporarily exceeds Coriolis force. Coriolis force cannot increase fast enough to balance the pressure gradient force. Thus the air swings slightly to the north across the contours. As the air leaves the jet max, the pressure gradient force decreases. Coriolis force temporarily exceeds the pressure gradient and the air swings slightly to the south, across the contours. This causes convergence (regions of downward-moving air from the surface through the jet stream level) in the northwest and southeast quadrants of the jet and divergence (regions of upward-moving air) in the southwest and northeast quadrants. The southwest quadrant is usually the more active because of greater temperature differences.

At the surface, deepening storms develop below the strongest divergence in the northeast quadrant. In the spring, summer, and early fall, one of these quadrants may be all that's required to trigger severe thunderstorms and tornadoes. These are mechanisms that forecasters look for in the development of the convective outlook and other severe weather products. Beneath the area of strongest convergence, the northwest quadrant, anticyclones build.

As mentioned, divergence aloft develops when contours diverge or move apart, as seen on the 300-mb chart. This causes surface convergence and increased cyclonic vorticity. Surface lows can develop in this way. A perfect example occurred one afternoon with scattered thunderstorms forecast for northern California, northern Nevada, and eastern Oregon. No weather systems were depicted on either the surface analysis or the 500-mb chart. However, thunderstorms did occur, right along a line of difluence. The

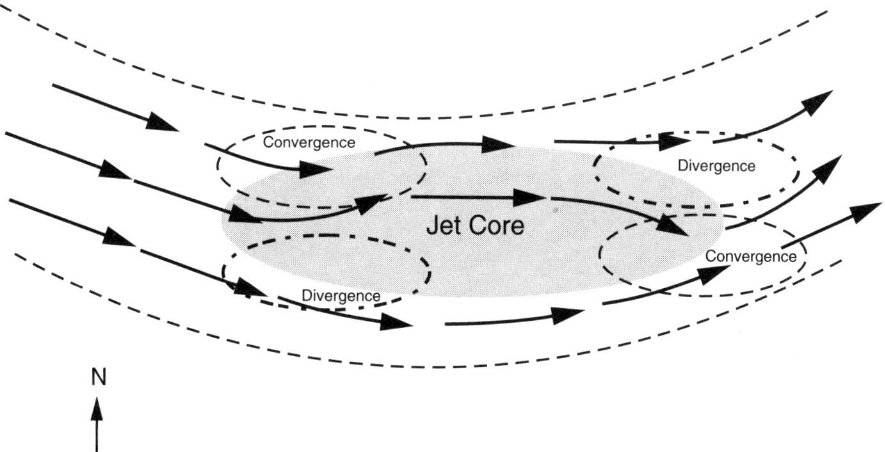

Fig. 7-9. *Areas of convergence and divergence are also associated with the jet stream.*

difluence caused just enough surface convergence, along with a moist, unstable air mass, to trigger thunderstorms.

HURRICANES

Hurricanes produce just about every kind of nasty weather extending over thousands of square miles. Figure 7-10 is an infrared satellite photo of hurricane Gilbert on September 11, 1988. A few days later, winds were reported to 140 knots, pressure in the eye was 26.66 in. Hg, tropical storm-force winds extended over a diameter of 450 miles, and squalls with rainfall of 5 to 10 inches accompanied the storm. The NWS hurricane advisory: "Gilbert may still increase a little more in strength."

Before we proceed, it might be interesting to see how hurricanes get their names. It appears that the practice of giving Atlantic hurricanes women's names began with writer George Stewart in his book *Storm,* published in 1941. A character in the book was a Weather Bureau meteorologist who used this method as he tracked these storms.

During World War II, Army and Navy weather types tracked storms over the Pacific as part of the war effort. Using women's names to communicate storm information was short, quick, and less-confusing than the system that had been used in the past. This practice continued after the war and was applied to Atlantic hurricanes.

Fig. 7-10. *Hurricanes produce just about every kind of nasty weather, extending over thousands of square miles.*

Names are alphabetically selected in advance and applied to successive seasonal tropical storms, starting with "A" and proceeding through the alphabet. Names selected are short, easily pronounced, quickly recognized, and easy to remember. A similar procedure is used for Pacific storms.

In the 1980s, the practice of using only women's names was changed. Tropical storms are still named alphabetically, with "A" the first storm of the season, however, male (e.g., Gilbert) and female names are used alternately.

Tropical cyclone is a general term applied to any low pressure area that originates over tropical oceans. (*Cyclone* comes from the Greek "kyklon," which refers to the coil of a snake. *Hurricane* does not have a clear origin, but it seems to have been used by the natives of the West Indies or Central America based on the word for "great wind.")

Tropical storms typically develop during the mid- to late-summer season. Principle hurricane months are August, September, and October, with most occurring in September. However, early-season hurricanes can develop during May, June, or July. Hurricanes that affect North America evolve in the warm tropical waters of the Atlantic, Caribbean, and Gulf of Mexico, and off the west coast of Mexico.

Tropical cyclones are classified according to their intensity based on average one-minute windspeeds. Wind gusts in these storms may be as much as 50% greater than average one-minute windspeeds. Tropical cyclones are internationally classified as

- *Tropical depression*—highest sustained winds up to 34 knots
- *Tropical storm*—highest sustained winds 35 to 64 knots
- *Hurricane*—highest sustained winds 65 knots or more

Hurricanes are further classified by intensity using the Saffir-Simpson scale of hurricane intensity, shown in Table 7-1.

Tropical cyclones develop under optimum sea surface temperature and weather systems that produce low-level convergence and cyclonic windshear. They favor tropical, easterly waves, troughs aloft, and areas of converging northeast and southeast trade winds along the intertropical convergence zone.

The low-level convergence associated with these systems, by itself, does not support the development of a tropical cyclone. The system must also have horizontal outflow—divergence—at high troposphere levels. Recall the earlier discussion of high pressure over low pressure. At high levels, there is anticyclonic flow. This combination creates a "chimney" in which air is forced upward, causing clouds and precipitation. Condensation releases large quantities of latent heat, which raises the temperature of the system and accelerates upward motion. The increased temperature lowers surface pressure (recall our discussion of surface warm-air advection) which increases low-level convergence. This draws more moisture-laden air into the system. When these chain-reaction events continue, a huge vortex is generated, which may culminate in hurricane force winds.

Tropical cyclones usually originate between 5° and 20° latitude. Tropical cyclones are unlikely within 5° of the equator because Coriolis force is so small, cyclonic circulation cannot develop. Winds flow directly into an equatorial low, and it rapidly fills.

Table 7-1. Hurricane Intensity.

Category	Central pressure (mb)	Storm surge (feet)	Mean wind (knots)
1. Weak	>980	4–5	64–83
2. Moderate	965–969	6–8	84–96
3. Strong	945–964	9–12	97–113
4. Very Strong	920–944	13–18	114–135
5. Devastating	<920	>18	>135

Tropical cyclones in the northern hemisphere usually move in a direction between west and northwest while in low latitudes. As storms move toward midlatitude, they come under the influence of the prevailing westerlies. Thus a storm may move very erratically, reverse course, or even circle. As the prevailing westerlies become dominant, storms recurve toward the north, then to the northeast, and finally to the east-northeast as they reach well into midlatitudes.

If a storm tracks along a coastline or over open sea, it gives up, slowly unleashing its devastation far from tropical regions. However, if the storm moves inland, it weakens due to surface friction and loss of its moisture source. Like the process of air mass modification, as storms curve toward the north or east, they begin to lose their tropical characteristics and acquire characteristics more like low pressure areas in midlatitudes. Its strength weakens as cooler air flows into the storm.

While developing, the cyclone has the characteristics of a circular area of broken to overcast, multilayered clouds. Numerous rainshowers and thunderstorms are embedded in these clouds. Coverage ranges from scattered to almost solid.

As cyclonic flow increases, thunderstorms and rain showers form into broken or solid bands paralleling the wind flow that spirals into the center of the storm. These spiral rain bands frequently are seen on radar and satellite images. In both Fig. 7-10 and Fig. 7-11, these cloud bands can easily be seen. All the hazards of weather associated with thunderstorm, including tornadoes, are present. Between the bands weather is less severe, but like upper-level lows, it can "sucker" an unwary pilot.

The "eye" usually forms in the tropical-storm stage and continues through the hurricane stage. Near the top of the thunderstorms, the air is relatively dry. Losing its moisture, the air begins to flow outward away from the center, in a diverging anticyclonic flow. This flow extends several hundred miles from the eye. The air begins to sink and warm as it reaches the limit of the storm, which results in clear skies outside the storm.

Surrounding the eye is a wall of clouds that may extend above 50,000 feet. This "wall cloud" contains torrential rains and the strongest winds in the storm. As a result of eye-wall thunderstorms, the air warms from the release of latent heat. This initiates a downward motion in the eye, which helps account for the absence of weather at the storm's center. In the eye, skies are cloud-free, flight conditions are smooth, and winds are comparatively light. The average diameter of the eye is between 15 and 20 miles, although sometimes smaller or larger.

Occasionally, eastern Pacific hurricanes reach southern California. Figure 7-11 shows a Pacific hurricane off Baja California. Since the energy comes from the ocean, these hurricanes dissipate rapidly over land. The moist, unstable remnants of these storms can be carried north and inland to affect central California and the southern part of the intermountain region. In Fig. 7-11, moisture resulting in thunderstorms can be seen over the Baja peninsula and the southern portion of the Southern California deserts. Analysis of these southerly disturbances was first studied in the 1930s. Because they sometimes approached from the southeast, they were called "Sonoran" storms, after the Mexican state of that name.

The position of tropical cyclones can be found on surface and constant pressure charts. Their forecast position appears on low- and high-altitude significant weather prognosis charts. Additionally, the NWS issues tropical storm and hurricane alerts. These alerts contain the plotted position of the storms, their intensity, forecast movement, and intensity trend.

PUTTING IT TOGETHER

A theme of this chapter has been how upper-level weather systems modify and direct surface weather. We've seen how they intensify or stabilize conditions at the surface, trigger

Fig. 7-11. *The moist, unstable remnants of Pacific hurricanes can be carried north and inland to affect central California and the southern part of the intermountain region.*

thunderstorms, and enhance or retard the intensity of frontal boundaries. At this point we have enough knowledge to put our three-dimensional atmosphere together. The following discussion will help us understand the development and decay of weather systems at the surface.

In Chapter 6 we discussed frontogenesis and frontolysis strictly from the perspective of events occurring at the surface. Now we are ready to include the influence of upper-level systems.

Let's begin by superimposing a longwave trough over the stationary front in Fig. 6-1, which is illustrated in Fig. 7-12. Colder air is located on the north, polar side of illustration and warmer air on the south, tropical side. Winds aloft are strong, which results in a strong windshear between the surface and upper levels. Now let's move a shortwave trough through the area, causing instability as warmer air rises and colder air subsides—vertical motion. Cyclonic circulation is enhanced by horizontal and vertical air motion.

As a result of the shortwave, the flow aloft develops into an area of converging air behind the shortwave. An area of diverging air forms ahead of the shortwave. At the surface, pressures change and windspeed increases. If there is no surface front, this mechanism can start the chain of events that brings air masses of different properties together. Presto—a front is born.

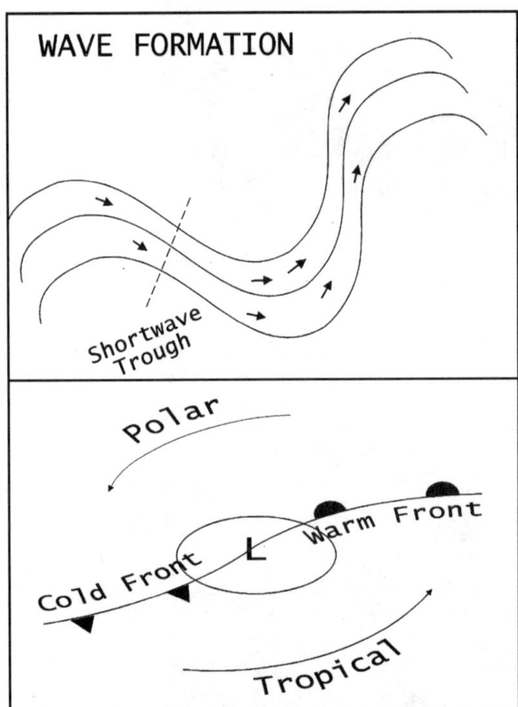

Fig. 7-12. *A three-dimensional picture of the atmosphere can be obtained by superimposing upper-level flow over surface conditions.*

As the converging surface air begins to spin, cold air flows southward and warm air northward—cyclonic flow. The stationary front now develops into distinct cold and warm fronts. Warm- and cold-air advection occurs below the 500-mb level.

Above the 500-mb level, cold-air advection occurs behind the cold front, bringing cold air into the trough. The cold-air advection makes the air more dense and lowers the height of the air column between the surface and the 500-mb level. Pressure in the trough lowers, and the trough deepens. Ahead of the trough, warm-air advection occurs, which has the effect of raising the height of the 500-mb surface, increasing pressure and the strengthening ridge. The result of the differential temperature advection is to intensify the strength of the developing cyclone. As condensation in the ascending air releases latent heat, the system strengthens even more. The result is the development of a midlatitude cyclone. Typically the most intense situation occurs when the upper-level trough is one-quarter wave length behind the surface cold front. Under these conditions, all the vertical motion producers become cumulative. The resulting weather patterns are those described in the life cycle of a front in Chapter 6.

Eventually the supply of warm surface air is cut off. A pool of cold air, which has broken off from the main flow, often lies directly above the surface low. Temperature advection ceases, and the surface system gradually weakens. The upper low itself may persist for days as a cut-off low, bringing clouds and precipitation to a large area with no surface fronts.

STRATEGIES

As in any weather situation, the key to hazardous weather avoidance is knowledge. This knowledge consists of a thorough understanding of basic weather phenomena, a complete weather briefing just prior to departure, and frequent weather updates en route.

The pilot must consider a number of other items, not necessarily weather-related. These include terrain, time of day, alternates if the planned flight cannot be completed, and the pilot's physiological and psychological conditions.

Throughout the book we've related terrain to specific aviation weather hazards and flight situations. Pilots must include aircraft performance in their decisions, as well as weather conditions. Aircraft performance and equipment refer to density altitude, service ceiling, availability of supplemental oxygen, ice protection, and thunderstorm avoidance equipment. If the aircraft does not have the performance or equipment, don't go! Other factors are alternate landing sites and time of day. Risks are increased flying over mountainous terrain at night in low-performance, single-engine aircraft.

The pilot's physiological and psychological condition must also be considered. Proper rest and good mental condition are paramount. It makes no sense for a tired, hung-over pilot to attempt to take on weather, or any flying for that matter. But that's another book. Other factors include the pilot's training and experience. Pilots need to set their own minimums based on these factors. It has been said that the new instrument rating is a ticket to learn. But let's get back to the weather.

VFR pilots must be careful in dealing with upper-level weather systems, especially in the absence of a front. Typically, weather conditions will be VFR to marginal VFR at the surface. In fact, pilots are often able to operate VFR beneath ceilings with relatively good visibility, except in the vicinity of showers, during these conditions. Here's the rub. In mountainous areas, there is almost always mountain obscuration, which prevents or precludes VFR operation in these areas. Pilots have to be especially careful not to be caught in box canyons. An IFR pilot typically has to contend with low freezing levels, turbulence, and thunderstorms, which are often embedded. To safely operate in these areas, the aircraft must have sufficient performance and ice protection and thunderstorm detection equipment. If you don't, don't go!

Some upper-level weather systems develop clouds and weather in bands. Typically, aviation forecasts cannot take this phenomenon into account. Therefore, the forecaster covers the area with a "broad brush" approach. Upon seeing a clear area, a pilot may assume the weather has passed. However, the weather then deteriorates with the approach of the next band. With closed lows and hurricanes, expect weather in bands. The satellite picture often confirms or refutes the existence of bands.

Another factor associated with upper-level lows and troughs is an unstable air mass. The low or trough may move through an area, bringing clear skies behind. However, with surface heating, abundant moisture, and an unstable air mass, guess what comes next? You got it—thunderstorms! Usually the forecast has this pretty well in hand, although, by now, we should be able to anticipate such phenomena.

A good source of severe weather information are convective SIGMETs (WST), alert weather watches (AWW), and the convective outlook (AC). These products contain information on the atmosphere that relates to severe weather. By now we should be able to understand and interpret most or all of the forecasters' discussions. These products give us additional insight into the overall weather picture. A couple of words of caution. An FSS briefer does not normally provide the outlook portion of WSTs, which contain the discussions, except on request. The convective outlook is just that, an outlook. The AC can never be used as a substitute for appropriate aviation weather forecasts (convective SIGMETs, area forecasts, and terminal forecasts).

The jet stream has its hazards of turbulence and strong winds. We have already discussed its significance and strategies. Hazards associated with vorticity are those produced by vertical motion mechanisms. The mention of these phenomena in forecasts or their appearance on charts should alert us to the fact that vertical motion is, or will be, occurring. En route, pilots should watch for unforecast windspeeds or shifts or temperature changes. These changes can signal a change in the weather pattern, which should cue the pilot to obtain additional information, such as current weather reports and revised forecasts, not to mention that a pirep alerts forecasters, briefers, and other pilots to a potentially hazardous, unforecast change in the weather.

Like thunderstorms, the key to the hurricane hazard is avoidance. Recall that thunderstorm tops associated with tropical cyclones frequently exceed 50,000 feet. Wind in a typical hurricane are strongest at low levels, decreasing with altitude. However, winds in excess of 100 knots at 18,000 feet are not uncommon. Severe to extreme turbulence and severe low-level windshear are to be expected. Turbulence increases in intensity in spiral rain bands and becomes most violent in the wall cloud surrounding the eye. Severe icing can be expected above the freezing level, which may be in the lower flight levels. Altimeter errors, due to the extreme low pressure, may be as much as 2000 feet. Flying to the right of the storm, a pilot can take advantage of a tailwind; to the left, the pilot will encounter the strongest headwinds, increasing fuel consumption and prolonging the risk.

The crew of the record-setting *Voyager* around-the-world flight in 1986 used this technique. In the western Pacific, off the Philippine Islands, they altered their flight plan to take advantage of the tailwinds north of a tropical storm. They had to be careful because the airplane was not constructed to withstand a significant amount of turbulence.

With our knowledge of the three-dimensional atmosphere, we should be able to put together the whole picture. By understanding why some weather systems are benign and others severe, and how they are modified, and integrating this knowledge with the preflight weather briefing and updates en route, we should be able to make intelligent, safe weather decisions. But this is only half of the safety equation. We have reviewed hundreds of accidents and weather scenarios. Most frequently an accident occurs because the pilot fails to obtain complete and accurate information, attempts to exceed aircraft performance, or simply continues flight beyond his or her own capability or below safe minimums.

Glossary

absolute instability A condition of the atmosphere in which vertical displacement is spontaneous, whether saturated or unsaturated.

absolute stability A condition of the atmosphere that resists vertical displacement whether a parcel is saturated or unsaturated.

advection The horizontal transport of an atmospheric property.

air mass A widespread body of air with homogeneous properties.

air to air visibility The visibility aloft between two aircraft or the aircraft and clouds.

air to ground visibility See slant-range visibility.

AIRMET bulletin An inflight weather advisory program intended to provide advance notice of potentially hazardous weather.

altimeter stetting The altimeter correction that allows the pilot to adjust the instrument for nonstandard pressure.

arc cloud An arc-shaped line of convective clouds often observed in satellite imagery moving away from a dissipating thunderstorm area.

atmospheric phenomena As reported on METAR, atmospheric phenomena is weather occurring at the station and any obstructions to vision. Obstructions to vision are only reported when the prevailing visibility is less than 7 miles.

Automated weather observing system (AWOS) A computerized system that measures sky conditions, visibility, precipitation, temperature, dewpoint, wind, and altimeter setting. It has a voice synthesizer to report minute-by-minute observations over radio frequencies, telephone lines, or local displays.

Automatic terminal information service (ATIS) A recorded service provided at tower-controlled airports to provide the pilot with weather, traffic, and takeoff and landing information.

backscatter device An instrument used to measure surface visibility at an automated weather-observing station.

Bjerknes, Vilhelm Norwegian meteorologist who developed the polar front theory at the beginning of the twentieth century.

blocking high An upper-level area of high pressure that blocks approaching weather systems. See *omega block.*

Celsius A temperature scale in which 0° is the melting point of ice and 100° is the boiling point of water.

Chinook An American Indian name given to a foehn wind on the eastern side of the Rockies.

closed low An area of low pressure aloft completely surrounded by a contour.

cold-core low A low pressure area that intensifies aloft. When this type of low contains closed contours at the 200-mb level, its movements tend to be slow and erratic.

conditional instability A condition of the atmosphere in which a parcel spontaneously rises as a result of it becoming saturated when forced upward.

conduction The process of transferring energy by means of physical contact.

confluence A region in which streamlines converge. The speed of the horizontal flow often increases where there is confluence. It is the upper-level equivalent of surface convergence.

constant pressure surface A surface where atmospheric pressure is equal; the height of the surface changes with pressure changes, but the pressure itself remains constant.

convection The vertical transport of an atmospheric property.

convergence An inward flow or squeezing of the air.

coordinated universal time (UTC) Formerly Greenwich mean time, also known as Z or Zulu time; the international time standard.

Coriolis force A force that deflects the wind, causing a rotational flow in weather systems.

Coriolis, Gaspard de A French mathematician who in 1835 developed the theory of an apparent deflection force produced by angular rotation.

cumuliform Clouds that are characterized by vertical development in the form of rising mounds, domes, or towers, and an unstable air mass.

cut-off low See *closed low.*

cyclone A closed, counterclockwise (in the northern hemisphere) circulation of low pressure.

cyclongenesis The development or strengthening of a cyclone.

cyclonic Having a counterclockwise rotation in the northern hemisphere, associated with the circulation around a cyclone or low pressure area.

deformation zone An area within the atmospheric circulation where air parcels contract in one direction and elongate in the perpendicular direction. The narrow zone along the axis of elongation is called the deformation zone.

density The weight of air per unit volume.

density altitude Density altitude is pressure altitude corrected for nonstandard temperature.

dewpoint The temperature to which air must be cooled, water vapor remaining constant, to become saturated.

dewpoint front See *dry line.*

diabatic A process that involves the exchange of heat with an external source, or non-adiabatic; the loss may occur through radiation resulting in fog or low clouds or conduction through contact with a cold surface.

difluence The spreading apart of adjacent streamlines. The speed of horizontal flow often decreases with a difluent zone. It is the upper-air equivalent of surface divergence and activates or perpetuates thunderstorm development.

Direct user access terminal A computer terminal at which pilots can directly access meteorological and aeronautical information, plus file a flight plan without the assistance of an FSS.

divergence A downward flow of air—subsidence; the opposite of convergence.

divergence Subsiding air spreads at the surface. Divergence is a downward motion producer that tends to stabilize the atmosphere near the surface.

drainage wind A wind directed down the slope of an incline and caused by greater air density near the slope than at the same levels some distance horizontally from the slope.

dry adiabatic lapse rate The rate at which unsaturated air cools or warms when forced upward or downward (3°C per 1000 feet).

dryline An area within an air mass that has little temperature gradient but significant differences in moisture. The boundary between the dry and moist air produces a lifting mechanism. Although not a true front, it has the potential to produce hazardous weather. It is also known as a dewpoint front.

enhanced infrared (IR) imagery A process by which infrared imagery is enhanced to provide increased contrast between features to simplify interpretation. This is done by assigning specific shades of gray to specific temperature ranges.

eye An area of clear skies that develops in the center of a tropical storm.

eye wall The area of thunderstorms that surrounds the eye of a tropical storm.

Fahrenheit A temperature scale in which 32° is the melting point of ice and 212° the boiling point of water.

fall streaks Ice crystals or snowflakes falling from high clouds into dry air, where they sublimate directly from a solid to a gas.

flight level Altitudes flown by the pilot with the altimeter set to 29.92 in. Hg; in the United States altitudes flown above 17.999 ft.

foehn wind A warm, dry wind on the lee side of a mountain range.

front A boundary between air masses of different temperatures, moisture, and wind.

frontal zone See *front.*

frontogenesis The process by which frontal systems are formed.

frontolysis The process of frontal system dissipation.

general circulation See *planetary scale.*

global circulation See **planetary scale.**

GOES Geostationary operational environmental satellites for North America; they are normally located about 22,000 nm above the equator at 75° W and 135° W. The satellites provide half-hourly visible and infrared imagery.

gravity wind See *drainage wind.*

Great Basin The area between the Rockies and Sierra Nevada mountains, consisting of southeastern Oregon, southern Idaho, western Utah, and Nevada.

gust front A low-level windshift line created by the downdrafts associated with thunderstorms and other convective clouds. Acting like a front, these features might produce strong gustiness, pressure rises, and low-level windshear.

Hadley cell See *Hadley, George.*

Hadley, George The first person, in 1735, to propose a direct thermally driven and zonally symmetric circulation as an explanation for the trade winds.

heat The total energy of the motion of molecules with the ability to do work.

heat burst A rapid, but brief, temperature jump associated with a thunderstorm.

hectopascal (hPa) The international unit of atmospheric pressure, equivalent to the millibar (mb).

high Area of high pressure completely surrounded by lower pressure

inches of mercury For aviation purposes, we commonly relate atmospheric pressure to inches of mercury (in. Hg), which is altimeter setting.

indicated airspeed The airspeed reading of the airspeed indicator uncorrected for temperature and pressure.

indicated altitude The altitude reading of the altimeter with the local altimeter setting in the altimeter setting window.

inflight visibility See *air to air visibility.*

infrared Satellite imagery measuring the relative temperature of clouds or the earth's surface.

instrument flight rules FARs that govern flight in instrument meteorological conditions; flight by reference to aircraft instruments.

intermountain region The area of the western United States, west of the Rocky Mountains and east of the Sierra Nevada Mountains, which includes Idaho and Arizona.

International standard atmosphere (ISA) A standard reference of temperature and pressure with a lapse rate of approximately 2°C per 1000 feet.

intertropical convergence zone (ITCZ) The convergence zone between the northeast trades and southeast trades near the equator.

inversion A lapse rate where temperature increases with altitude.

isobars Lines connecting equal values of surface pressure.

isotachs Used on charts and graphs, lines of equal windspeed.

isothermal A constant lapse rate; temperature remains constant with altitude.

isotherms Used on charts and graphs, lines of equal temperature.

jet streaks See *jet stream.*

jet stream A segmented band of strong winds that occur in breaks in the tropopause.

jetlets See *jet stream.*

katabatic wind Any wind blowing down an incline.

Kollsman, Paul A German-born aeronautical engineer who invented the method to correct the altimeter for nonstandard pressure in 1928.

lapse rate The decrease of an atmospheric variable with height, usually temperature.

lee-side low Low pressure areas that develop east of mountain barriers.

level of free convection (LFC) The point at which a parcel becomes saturated and upward movement becomes spontaneous.

lifted condensation level (LCL) The level or altitude at which a lifted parcel becomes saturated.

longwave see *Rossby waves.*

low An area of low pressure completely surrounded by higher pressure.

low-level windshear Windshear that occurs within 2000 feet of the surface.

mesolow Also known as *mesocyclone*; a small area of low pressure within a severe thunderstorm. Tornadoes can develop within the vortex.

mesoscale See *synoptic scale,* small-scale meteorological phenomena that can range in size from that of a single thunderstorm to an area the size of the state of Oklahoma.

mesoscale convective complexes (MCC) A large, organized convective weather system that can cover an area the size of several states.

METAR Meteorological aviation routine surface weather report.

microburst A small-scale, severe storm downburst less than 2½ miles across. Reaching the ground, the burst continues as an expanding outflow, producing severe windshear.

microscale See *subsynoptic scale.*

millibar For aviation purposes, we commonly relate atmospheric pressure to inches of mercury or in millibars (mb).

mixing ratio The ratio of water vapor to dry air, expressed in grams of water vapor per kilogram of dry air.

moisture convergence An objective analysis field combining wind flow convergence and moisture advection. Under certain circumstances, this field is useful for forecasting areas of thunderstorm development.

negative tilt Troughs with an axis in the horizontal plane tilting from northwest to southeast. These systems tend to cause more weather in California because they bring in warm, moist air.

negative vorticity advection Area of low values of vorticity, producing downward vertical motion.

neutral stability An atmospheric condition in which a parcel remains at rest after it is displaced, even when the displacing force ceases.

omega block A blocking high that on weather charts resembles the Greek letter *omega.*

orographic The effects caused by terrain, especially mountains, on the weather.

outflow boundary A surface boundary left by the horizontal spreading of thunderstorm-cooled air; often the lifting mechanism needed to generate new thunderstorms.

overrunning A condition in which air flow from one air mass is moving over another air mass of greater density. The term usually applies to warmer air flowing over cooler air, as in a warm frontal situation, implying a lifting mechanism that can trigger convection in unstable air.

parcel A small volume of air arbitrarily selected for study; it retains its composition and does not mix with the surrounding air.

Pascal, Blaise A French philosopher and mathematician (1623–1662) for whom the international unit of atmospheric pressure, the hectopascal (hPa), is named.

planetary boundary layer The frictional layer between the surface and the atmosphere.

planetary scale The jet stream, subtropical high, polar front, and intertropical convergence zone.

polar front A semipermanent, semicontinuous front separating air masses of tropical and polar origin.

polar jet The jet stream located at the break between the polar tropopause and subtropical tropopause.

positive tilt Troughs with an axis in the horizontal plane tilting from northeast to southwest.

positive vorticity advection Areas of high positive values of vorticity, producing upward vertical motion.

pressure Force per unit area.

pressure altitude The altitude above the mean sea level constant pressure surface; indicated altitude with the altimeter set to 29.92 in. Hg.

prevailing visibility Visibility reported in manual observations; the greatest distance that can be seen throughout at least half the horizon circle, which need not be continuous.

QFE Altimeter setting used so that the altimeter reads zero when the aircraft is on the ground.

QNE Altimeter setting used to obtain pressure altitude.

QNH Altimeter setting used to obtain indicated altitude.

radiation The process of transferring energy through space without the aid of a material medium.

relative humidity The ratio, expressed as a percentage, of water vapor present in the air compared to the maximum amount the air could hold at its present temperature.

ridge An elongated area of high pressure.

Rossby waves Longwave or global waves in the upper troposphere in midlatitudes that circle the earth.

Santa Ana The local name given to a foehn wind that occurs in late fall, winter, and early summer in Southern California.

saturated adiabatic lapse rate The rate at which saturated air cools or warms when forced upward or downward.

scud Shreds of small detached clouds moving rapidly below a solid deck or higher clouds.

severe thunderstorms A thunderstorm that produces winds of 50 knots or greater, or hail $\frac{3}{4}$ inches or greater.

shear axis An axis indicating maximum lateral change in wind direction, as in an elongated circulation. This lateral change or shear might be either cyclonic or anticyclonic.

shortwave A trough or ridge embedded in a longwave that moves through the longwave at the same speed as the overall wind flow.

showers Precipitation characterized by the suddenness with which it starts and stops and rapid changes of intensity.

SIGMET A significant meteorological advisory that warns of phenomena that affects all aircraft.

slant-range visibility The visibility between an aircraft in the air and objects on the ground.

Sonora storm Local name given to a storm that approaches Southern California from the southeast.

SPECI A special surface aviation weather report.

squall line An organized line of thunderstorms.

stratiform Clouds of extensive horizontal development and a stable air mass.

stratosphere The layer of the atmosphere above the tropopause.

streamlines A line that represents the wind flow pattern and is parallel to the instantaneous velocity. Streamlines indicate the trajectory of the flow.

subsynoptic scale Small, microscale events that often are not within the detection system currently available, such as tornadoes and microbursts.

sublimation The process by which ice changes directly to water vapor or water vapor directly to ice. The sublimation of ice to vapor is a cooling process and of water vapor to ice is a warming process.

subtropical jet The jet around 20° to 30° north latitude at approximately 39,000 feet, located in the break between the midlatitude and tropical tropopause.

supercooled Liquid water or water vapor that exists at temperatures below freezing.

surface visibility The visibility at and along the surface of the earth.

synoptic scale Mesoscale circulation consisting of highs and lows and troughs and ridges, surface and aloft.

TAF Terminal aerodrome forecast.

temperature A measurement of the average speed of molecules.

thermal high An area of high atmospheric pressure caused by the cooling of air by a cold surface. They remain relatively stationary over the cold ground.

thermal low An area of low atmospheric pressure caused by intense surface heating. They are common to the continental subtropics in summer, remain stationary, and cyclonic circulation is generally weak and diffuse.

Torricelli, Evangelista The Italian inventor of the mercurial barometer (1643).

transverse cirrus banding Irregularly spaced band-like cirrus clouds that form nearly perpendicular to a jet stream axis. They indicate turbulence associated with the jet.

tropical cyclone A general term applied to any low pressure area that originates over tropical oceans.

tropopause The boundary between the troposphere and the stratosphere.

troposphere The lower layer of the earth's atmosphere containing about three-quarters of the atmosphere by weight.

trough An elongated area of low pressure.

true altitude Indicated airspeed corrected for temperature and pressure.

upslope The orographic effect of air moving up a slope, which tends to cool adiabatically.

virga Rain that evaporates before reaching the surface.

visual flight rules (VFR) FARs that govern flight in visual meteorological conditions; flight by reference to the natural horizon and surface.

vort lobe A contraction for *vorticity lobe.* It usually applies to the 500-mb level and identifies an area of relatively higher values of vorticity. It is synonymous with short-wave trough or upper-level impulse. Generally speaking, there is rising air ahead of the vort lobe and sinking air behind.

vort max A contraction for *vorticity maximum.* It usually applies to the 500-mb level and refers to a point along a vorticity lobe where the absolute vorticity reaches a maximum value.

vorticity A mathematical term that refers to the tendency of the air to spin; a vertical motion producer.

wall cloud See *eye wall.*

warm-air advection A condition in the atmosphere characterized by air flowing from a relatively warmer area to a cooler area. It is often accompanied by upward vertical motion that in the presence of sufficient instability leads to thunderstorm development.

warm-core low An area of low pressure that is warmer at its center than at its periphery. Thermal lows and tropical cyclones are examples.

water vapor The invisible water molecules suspended in the air.

wave A pattern of ridges and troughs in the horizontal flow as depicted on upper-level charts. At the surface, a wave is characterized by a break along a frontal boundary. A center of low pressure is frequently located at the apex of the wave.

wind field Winds plotted at a specified level in the atmosphere are referred to as a wind field (surface, 5000 feet, 300-mb, etc.). Wind fields show areas of convergence, divergence, and advection, which provide meteorologists with a valuable forecast tool.

windshear Any rapid change in wind direction or velocity. Low-level windshear (LLWS) is generally shear that occurs within about 2000 feet of the surface. LLWS is classified as severe when a rapid change in wind direction or velocity causes an airspeed change greater than 15 knots or vertical speed change greater than 500 feet per minute.

zonal flow A wind flow that is generally in a west-to-east direction.

Appendix A
Charts and graphs

THIS APPENDIX CONTAINS CELSIUS-TO-FAHRENHEIT AND METERS-TO-statute-miles conversions charts and approximate density altitude calculation charts and graphs. An example of an approximate density altitude calculation is also included.

Recall from Chapter 1 that density altitude is based on field elevation corrected for nonstandard pressure and temperature. Given the following data, calculate approximate density altitude.

Field elevation: 4000 feet
Altimeter setting: 29.73 in. Hg
Temperature: 75°F

Since our density altitude graph requires temperature in degrees Celsius and pressure altitude, these will be our first conversions.

From the temperature conversion chart, we determine that 75°F equals 24°C. The pressure altitude chart allows us to correct for nonstandard pressure. Note that the correction factor for 29.92 in. Hg (standard pressure) is zero. Pressure higher than standard results in a pressure altitude lower than field elevation; pressure lower than standard causes pressure altitude to be higher than field elevation. Therefore, add positive values to field elevation and subtract negative values.

Pressure altitude: 4200 feet
Temperature: 24°C

Refer to the density altitude graph.

1. Locate the temperature (24°C) on the bottom horizontal scale.
2. Track up the graph to the intersection of the 4200-foot pressure altitude (diagonal) line, right vertical scale (illustrated by the vertical dashed line on the density altitude graph).
3. Track left to the density altitude scale, left vertical scale; read the approximate density altitude 6200 feet (illustrated by the horizontal dashed line on the density altitude graph).

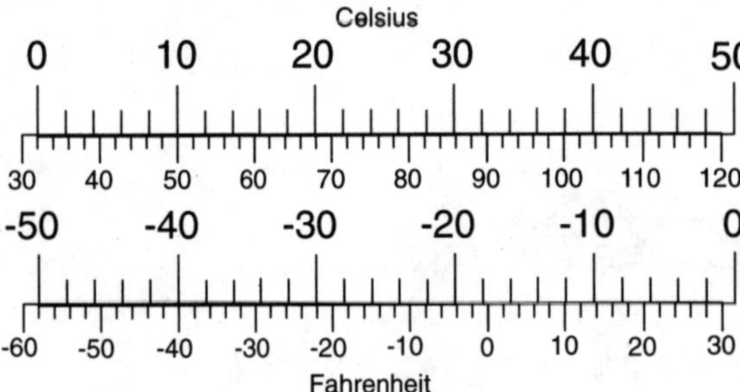

Visibility in Meters:

Meters	Statute ML	Meters	Statute ML
0000	0	2200	1 3/8
0200	1/16	2400	1 1/2
0300	1/8	2600	1 5/8
0400	3/16	2800	1 3/4
0500	5/16	3000	1 7/8
0600	3/8	3200	2
0800	1/2	3600	2 1/4
1000	5/8	4000	2 1/2
1200	3/4	4800	3
1400	7/8	6000	4
1600	1	8000	5
1800	1 1/8	9000	6
2000	1 1/4	9999	6+

Fig. A-1. *Temperatures in METAR reports are degrees Celsius; visibility on military TAFs are meters.*

PRESSURE ALTITUDE							
ALSTG	COR	ALSTG	COR	ALSTG	COR	ALSTG	COR
29.10 29.14	+800	29.50 29.54	+400	29.90 29.94	0	30.30 30.34	-400
29.15 29.19	+750	29.55 29.59	+350	29.95 29.99	-50	30.35 30.39	-450
29.20 29.24	+700	29.60 29.64	+300	30.00 30.04	-100	30.40 30.44	-500
29.25 29.29	+650	29.65 29.69	+250	30.05 30.09	-150	30.45 30.49	-550
29.30 29.34	+600	29.70 29.74	+200	30.10 30.14	-200	30.50 30.54	-600
29.35 29.39	+550	29.75 29.79	+150	30.15 30.19	-250	30.55 30.59	-650
29.40 29.49	+500	29.80 29.84	+100	30.20 30.24	-300	30.60 30.64	-700
29.45 29.49	+450	29.85 29.89	+50	30.25 30.29	-350	30.65 30.70	-750

Fig. A-2. *Pressure altitude is required to calculate density altitude.*

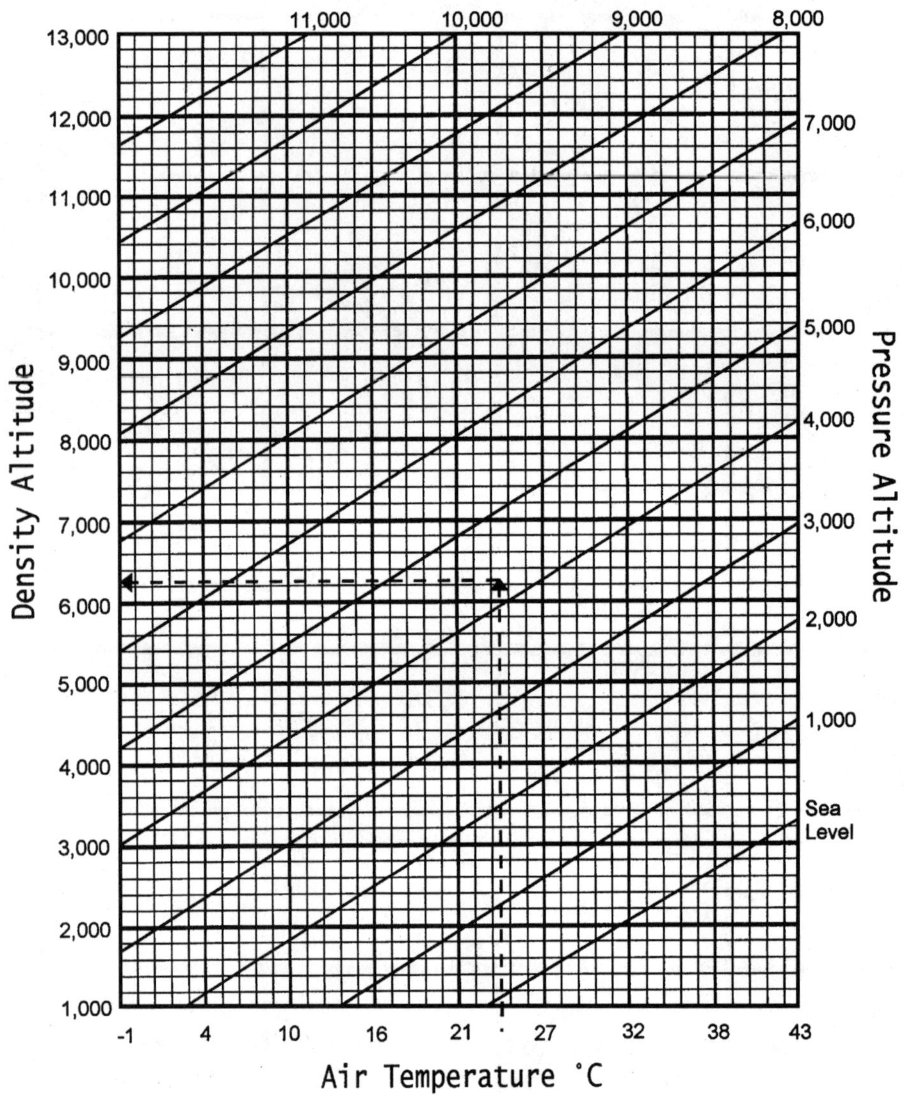

Fig. A-3. *Density altitude is pressure altitude corrected for nonstandard temperature.*

Appendix B METAR/TAF codes

METAR (METEOROLOGICAL AVIATION ROUTINE) REPLACED THE North American SA (surface aviation) code, and TAF (terminal aerodrome forecast) replaced the domestic FT (terminal forecast) in the United States on July 1, 1996. Mexico converted in 1995 and Canada on June 3, 1996. U.S. military terminal forecasts and NWS international airport forecasts have used the TAF code and format for years. Ostensibly, METAR will standardize surface aviation reports with the rest of the world. With METAR and TAF, individual member countries are allowed to change certain items in the reports. For example, the United States will continue to use the current units of measurement rather than metric, except for temperatures. This is one reason for the format used in the United States compared to those of other countries.

In the METAR and TAF codes, the letter "M" means minus or less than, and the letter "P" means plus or more than.

Essentially, METAR and TAF contain the same information as the SA and FT codes, but in a different order and slightly different format. Interpretation of various elements remains the same.

METAR (METEOROLOGICAL AVIATION ROUTINE)

METAR reports contain the following sequence of elements:

- Type of report
- Station identifier
- Date and time of report
- Report modifier, if required
- Wind
- Visibility
- Weather and obstructions to visibility
- Sky condition
- Temperature and dewpoint
- Altimeter setting
- Remarks

When elements in METAR are missing, they are omitted.

Type of Report

There are two types of reports. METAR is a routine observation. SPECI is a special METAR weather report.

Station Identifier

METAR uses standard four-letter ICAO (International Civil Aviation Organization) identifiers. For the continental United States, this identifier is the domestic identifier prefixed with the letter "K." For example, Newark "KEWR" and Philadelphia is "KPHL."

Date/Time Group

The time of observation is transmitted as a six-digit date/time group appended with a "Z" to denote coordinated universal time (UTC). The first two digits represent the day of the month; the last four digits the time of observation.

Report Modifier

Two report modifiers may appear after the date/time group. AUTO indicates the report comes from an automated weather observation station that is not augmented. *Augmented* means someone is physically at the site monitoring the equipment. COR means the report was originally transmitted with an error and that has now been corrected.

Wind

Wind is reported as a two-minute prevailing wind direction and speed using a five-digit group, six if the speed exceeds 99 knots. The first three digits indicate the direction the wind is blowing from in relation to true north. The next two or three digits indicate speed. The units of measurement follow (KT knots, KMH kilometers per hour, MPS meters per second). In the United States, knots will continue to be used (34017KT—wind 340° at 17 knots). Gustiness is reported with a "G" (23020G30KT). If wind direction varies by 60° or more with a speed greater than 6 knots, a variable group separated by a "V" will follow the prevailing group (34017KT 310V010). A calm wind is reported as 00000KT. The contraction VRB indicates a variable wind direction. VRB may be included for winds less than 7 knots (VRB04KT). VRB may also be used in special cases at higher speeds, such as a wind shift as a thunderstorm passes over the station.

Visibility

In the United States, visibility, as before, is reported as prevailing, or its automated station equivalent, in statute miles "SM." For example, 1/2SM is ½ statute mile, 7SM is 7 statute miles, and 15SM is 15 statute miles. For automated stations, a visibility of less than ¼ mile is reported M1/4SM.

Runway visual range, when reported, uses the following format: the letter "R" followed by the runway number, a solidus "/," and the RVR in feet. For example: R29/2400FT (runway 29 visual range 2400 feet). The following codes will be added to the visual range as required:

- V—variability (R32R/1600V2400FT)
- M—less than (R22R/M1600FT)
- P—more than (R36/P6000FT)

Weather and Obstructions to Vision

Weather and obstructions to vision in the METAR code represent a significant change from SA reports. In METAR, weather and obstructions to vision are reported in the following format: intensity, proximity, descriptor, precipitation, obstruction to vision, and other.

Refer to the weather and obstructions to vision chart at the end of this appendix. The intensity of precipitation is indicated in the same way as on SAs, except it is located at the beginning of the group. It's important to remember that intensity refers to the precipitation, not the descriptor. For example, +TSRAGR is a thunderstorm with heavy rain and hail. A severe thunderstorm is indicated by surface winds of 50 knots or hail ¾ inch in diameter.

Proximity, or vicinity ("VC"), reports weather occurring in the vicinity of the airport. Vicinity is defined for precipitation as not occurring at the point of observation but within 10 statute miles of the station; when used to report any type of fog, vicinity means from between 5 and 10 statute miles of the station.

Descriptors apply to precipitation or obstructions to vision. Automated sites may use "UP" to report precipitation of unknown type. Fog (FG) is only reported when the visibility is less than ⅝ mile. With visibility between ⅝ and 6 miles, mist (BR) is used. In METAR, mist (BR) refers to an obstruction to vision, not precipitation.

Five categories of weather phenomena appear in the "other" group. A squall (SQ) is a strong wind characterized by a sudden onset, a duration of some minutes, and then a sudden decrease in speed. Sandstorms (SS) and duststorms (DS) are differentiated from blowing dust and blowing sand by visibility. Sandstorms and duststorms only appear when the visibility is less than $\frac{5}{8}$ of a mile. Dust or sand whirls (PO) are dust or sand raised by rapidly rotating columns of air. When well developed, they are dust devils.

Sky Condition

METAR sky condition reports represents another significant change in criteria and format. Thin layers are no longer reported. Format consists of the amount of cloud coverage, height agl of the layer, and, under certain conditions, cloud type (TCU and CB).

The amount of sky cover is reported in eighths. A clear sky is reported using the contraction SKC (sky clear).

- CLR—clear below 12,000 feet (automated reports)
- SKC—clear (no clouds)
- Few—few (less than $\frac{1}{8}$ to $\frac{2}{8}$ coverage)
- SCT—scattered ($\frac{3}{8}$ to $\frac{4}{8}$ coverage)
- BKN—broken ($\frac{5}{8}$ to $\frac{7}{8}$ coverage)
- OVC—overcast ($\frac{8}{8}$ coverage)

When a cloud layer develops below the point of observation, such as at Mt. Wilson above the Los Angles Basin, the layer is encoded OVC/// (an overcast layer with tops below the point of observation).

A partial obscuration is reported when between $\frac{1}{8}$ and $\frac{7}{8}$ of the sky is hidden by a surface-based obscuring phenomena. In the METAR code, a partial obscuration is indicated as FEW, SCT, or BKN on the surface (FEW000, between $\frac{1}{8}$ and $\frac{2}{8}$ of the sky obscured). The remark indicates it is a partial obscuration (...RMK FG SCT000). What else could it be? Well, technically, a layer with a base of less than 50 feet would be reported as "000." This is very unlikely. If the observer did indeed mean to report a layer with a base less than 50 feet, the remark "FG SCT000" would not appear. (At this writing the FAA and NWS are still trying to eliminate partial obscurations from U.S. METAR reporting criteria.)

A ceiling is not designated. For aviation purposes, the ceiling will remain the lowest broken or overcast layer or vertical visibility into a total obscuration except that a partial obscuration does not constitute a ceiling.

Cloud heights are reported with three digits, in hundreds of feet.

When towering cumulus TCU (cumulus congestus) or cumulonimbus CB are present, they will be included in the sky condition section of the report (BKN035TCU—towering cumulus clouds with a ceiling 3500 broken).

Vertical visibility into a total obscuration is reported using the letters "VV" followed by the vertical visibility in hundreds of feet (VV002—indefinite ceiling 200 feet).

Temperature, Dewpoint, and Altimeter

Temperature and dewpoint are reported using two digits in degrees Celsius. Temperatures below zero are prefixed with the letter "M." For example, 20/15—temperature 20°C and dewpoint 15°C—or 08/M03—temperature 8°C and dewpoint −3°C.

The altimeter, in inches of mercury, is reported in a four-digit format prefixed with the letter "A," for example, A2992—altimeter setting two nine point nine two inches of mercury.

Remarks

Remarks in METAR follow the contraction "RMK." Remarks are divided into automated, manual, and plain language and additive and automated maintenance date. Automated remarks indicate the type of automated station. Manual remarks supplement information in the body of reports. Plain language are remarks significant to other agencies, for example, FIRST or LAST observation of the day. Additive date is for the NWS and consists of climatological information, usually in a numerical code group. Automated maintenance data for automated reports consist of sensor outages and maintenance requirements.

The following are examples of METAR reports.

SPECI KEWR 172240Z AUTO 03545G60KT 0SM R04/4500FT +TSRA SQ VV000 23/21 A2989 RMK AO2 TS OVHD MOVG ESE FQT LTGICCCCG

Decode: Special report for Newark, NJ; taken on the 17th day of the month at 2240 UTC by a nonaugmented automated weather observation station; wind 030° at 45 knots with gust to 60 knots; visibility zero, runway 4 visual range 4500 feet in thunderstorm with heavy rain and squalls; indefinite ceiling zero; temperature 23°C, dewpoint 21°C; altimeter 29.89, remarks automated station with precipitation sensor, thunderstorm overhead moving east-southeast, frequent lightning in cloud, cloud to cloud, and cloud to ground.

METAR KPHL 172250Z COR 36018G24KT 1SM R27R/3000VP6000FT +TSRA VV006 22/20 A2983 RMK TSB25 ALQDS MOVG SE FQT LTGICCG RAB38 PRESFR

Decode: Routine hourly observation for Philadelphia, PA; taken on the 17th day of the month at 2250 UTC; the report has been corrected; wind 360° at 18 knots with gusts to 24 knots; visibility 1 statute mile, runway 27 right visual range variable between 3000 and more than 6000 feet; thunderstorm with heavy rain; indefinite ceiling 600 feet; temperature 22°C, dew point 20°C; altimeter setting 29.83; remarks thunderstorm began at 25 minutes past the present hour, thunderstorm all quadrants moving southeast with frequent lightning in cloud and cloud to ground, rain began 38 minutes past the present hour, the pressure is falling rapidly.

TAF (TERMINAL AERODROME FORECASTS)

According to sources within the National Weather Service, "a few of the changes seem to defy logic—in fact they are illogical." At first glance the new format appears confusing, but actually the basic forecast remains the same. TAFs are issued four times a day (0000Z, 0600Z, 1200Z, 1800Z) and are valid for 24 hours. Forecasts using TAF (World Meteorological Organization, WMO) codes are also issued by local base weather offices for many military locations. Military TAFs have different criteria than those issued by the NWS; therefore, some differences will occur.

TAFs are issued in the following format:

- Type
- Location
- Issuance time
- Valid time
- Forecast

There are two types of TAF issuance, a routine forecast issuance, "TAF," and an amended forecast, "TAF AMD." Like domestic FTs, TAFs may be corrected "COR" or routinely delayed "NIL." TAF location is identified by the four-letter ICAO station identifier. Issuance date and time consists of a six-digit group. The first two digits represent the day of the month and the last four digits, UTC issuance time. The valid period is a four digit group, usually 24 hours, in UTC. The body of the TAF uses the following format:

- Wind
- Visibility
- Weather
- Sky condition
- Windshear, when applicable

Like the METAR code, wind is forecast as a five- or six-digit group when considered significant to aviation. The contraction "KT" follows the wind forecast and denotes the units as knots. Gusts are noted by the letter "G."

Prevailing visibility up to and including 6 statute miles is forecast. Visibility greater than 6 miles is indicated by the letter "P" for plus (P6SM—visibility greater than 6 statute miles). Military and many international TAFs forecast visibility in meters. A conversion from meters to miles is contained in Appendix A.

Weather and obstructions to vision use the same format and codes as the METAR report. With no significant weather expected, the weather group is omitted. When significant weather is forecast but is expected to change in the future to no significant weather, the contraction "NSW" (no significant weather) appears. NSW does not appear when any of the following phenomena are expected to occur.

- Freezing precipitation
- Moderate or heavy precipitation
- Drifting or blowing dust, sand, or snow
- Duststorms or sandstorms
- Thunderstorms
- Squall

- Tornadoes
- Phenomena expected to cause a significant change in visibility

Sky condition uses the same format and contractions as METAR, that is, amount, height, cloud type, or vertical visibility. When cumulonimbus clouds are expected, "CB" is appended to the cloud layer. CB is the only cloud type forecast in TAFs.

Low-level windshear, when forecast, appears following sky condition on domestic TAFs, in the following format:

WS015/24035KT

- **WS:** windshear
- **015:** height in hundreds of feet (above ground level) of the windshear (1500 feet agl).
- **/24035:** wind direction and speed (knots) above the windshear (240° at 35 knots).

Codes that may appear on international and military TAFs are contained in the chart at the end of the appendix.

Conditional Terms

TAF conditional terms consists of temporary (TEMPO) conditions and probability (PROB) forecasts. TEMPO indicates that temporary conditions are expected to occur during the forecast period. TEMPO describes any condition expected to last for generally less than an hour at a time. The time during which the condition is expected to occur is indicated with a four-digit group giving beginning and ending time UTC. For example, SCT030 TEMPO 1923 BKN030 means 3000 scattered temporarily between 1900Z and 2300Z ceilings 3000 broken. A ceiling of 3000 broken is expected to exist for periods of less than one hour during the 1900Z to 2300Z time frame. TEMPO is equivalent to occasional (OCNL) on the old domestic FTs.

A PROB group indicates the probability of occurrence of thunderstorms or other precipitation events. PROB 40 indicates a 40% probability; PROB 30 indicates a 30% probability. This code is followed by a four-digit time group giving beginning and ending times. PROB40 is the equivalent to CHC, and PROB 30 is equivalent to SLT CHC on the old domestic FTs.

Forecast Change Groups

Forecast change groups consist of from (FM) followed by a time group (tttt)—FMtttt—and becoming (BECMG) followed by a time group (TTtt)—BECMG TTtt. The FMtttt group is used when a rapid change is expected, usually less than one hour. For example, BKN020 FM1630 SKC—before 1630Z ceiling 2000 broken, around 1630Z the sky condition will change to clear. The BECMG TTtt group indicates a more gradual change in conditions over a longer period of time, usually two hours. For example, 5SM HZ BKN 030 BECMG 0507 3SM BR OVC 020—before 0500Z visibility 5 in haze, ceiling 3000 broken, then during the period of 0500Z to 0700Z, conditions changing to visibility 3 miles in fog, ceiling 2000 overcast.

An example of a TAF for Seattle Tacoma International Airport follows:

TAF

KSEA 041045Z 1212 00000KT P6SM SCT015 OVC025 TEMPO 1214 -RA

BECMG 1314 24007KT SCT015 BKN025 TEMPO 1418 -SHRA BECMG

1718 SCT020 BKN035 PROB40 1801 -TSRA BECMG 0001 22006KT

SCT015 BKN035 TEMPO 0112 -SHRA=

This is the TAF for Seattle Tacoma (KSEA) issued on the fourth day of the month at 1045Z (041045Z). The forecast is valid from the fourth at 1200Z until the fifth at 1200Z (1212). At 12Z the surface winds are expected to be calm (00000KT). Visibility forecast more than 6 statute miles (P6SM). Sky condition 1500 scattered ceiling, 2500 overcast. Occasionally (TEMPO) between 1200Z and 1400Z (1214) light rain is expected. Prevailing conditions are expected to become (BECMG) between 1300Z and 1400Z (1314) wind 240 at 7 (visibility is implied to be more than 6 miles), 1500 scattered ceiling, 2500 broken. Occasionally between 1400Z and 1800Z light rain showers are expected. Between 1700Z and 1800Z (wind and visibility are implied to be 240 at 7 and more than 6 miles) 2000 scattered with ceilings of 3500 broken are forecast. There is a 40% chance (PROB40) of thunderstorms with light rain. Between 0000Z and 0100Z wind 220 at 6, visibility more than 6, 1500 scattered, ceiling 3500 broken are expected, occasionally between 0100Z and 1200Z light rain showers are forecast.

Following is an example of Tacoma's McCord Air Force Base forecast for the same time period as the previous example.

TCM 1212 20009KT 9999 SCT010 BKN020 OVC050 T12/13 540109 610703 QNH2980INS CIG020

BECMG 1314 20009KT 9999 SCT015 BKN035 OVC045 540109 620703 QNH2975INS CIG035

BECMG 1920 24007KT 9999 SCT020 BKN045 BKN150 T20/23 540009 620505 QNH2980INS CIG045

BECMG 0001 VRB05KT 9999 -SHRA SCT020 BKN040 OVC070 610505 QNH2983INS CIG040

BECMG 0405 VRB05KT 9999 NSW SCT020 BKN050 QNH2990INS

CIG050=

There are several major differences between domestic TAFs and those issued by the military and foreign governments. Visibility forecasts are in meters. In the example, the visibility through the forecast period is 9999—greater than 9000 meters. Sky condition forecasts do not use the summation principle, and the ceiling is specified (BKN015 OVC080 BKN200 CIG015). The altimeter setting is forecast (QNH2980INS—altimeter setting two nine point eight zero inches of mercury).

Maximum and minimum temperatures, icing, and turbulence are sometimes forecast. The temperature group follows sky conditions. For example, T12/13—minimum temperature 12°C at 1300Z; T20/23—maximum temperature 20°C at 2300Z. Turbulence and icing forecasts may appear in the format illustrated in the chart at the end of the appendix. In the example at 1200Z, 540109 appears. The 5 represents the turbulence group. The 4 indicates the turbulence intensity—moderate turbulence in cloud, infrequent. The next group, 010 shows the base of the turbulence layer height in hundreds of feet—1000 feet. The 9 represents the thickness of the turbulence layer in thousands of feet—9000 feet. Moderate turbulence is forecast between 1000 and 10,000 feet.

In the example at 1200Z, 610703 appears. The 6 represents the icing group. The 1 indicates the icing intensity—light icing. The 070 shows the base of the icing layer height in hundreds of feet—7000 feet msl. The 3 represents the thickness of the icing layer in thousands of feet—3000 feet. Light icing is forecast between 7000 and 10,000 feet. Since turbulence and icing forecasts are provided in the AIRMET bulletin and SIGMETs, U.S. domestic TAFs do not contain these phenomena.

Appendix B

Weather and Obstruction to Vision

Intensity	
-	Light
no symbol	Moderate
+	Heavy
Proximity	
VC	Vicinity
Descriptor	
TS	Thunderstorm
DR	Low Drifting
SH	Showers
MI	Shallow-*mince*
FZ	Freezing
BC	Patches-*banc*
BL	Blowing
PR	Partial
Precipitation	
RA	Rain
DZ	Drizzle
GR	Hail (<> than 1/4 in.)-*grêle*
GS	Small Hail/Snow Pellets-*grésil*
SN	Snow
PE	Ice Pellets
SG	Snow Grains
IC	Ice Crystals
UP	Precipitation (Automated Observation)
Obstructions to Vision	
FG	Fog (Visibility less than 5/8)
PY	Spray
BR	Mist (Visibility 5/8 to 6)-*brume*
SA	Sand
FU	Smoke-*fumée*
DU	Dust
HZ	Haze
VA	Volcanic Ash
	Other
SQ	Squall
SS	Sandstorm
DS	Duststorm
PO	Dust/Sand Whirls
FC	Funnel Cloud
+FC	Tornado
+FC	Waterspout

Fig. B-1. *Weather and obstruction to vision codes in METAR are derived from French as well as English words.*

Terminal Aerodrome Forecasts (TAF)

Temperature: TT_1T_1/tt
T - Temperature Group
T_1T_1 - Temperature in Celsius
tt - Time UTC

Turbulence: 5ihhhd
5 - Turbulence Group
i - Turbulence Intensity
hhh - Base Height hundreds of feet
d - Thickness in thousands of feet

Icing: 6ihhhd
6 - Icing Group
i - Icing Intensity
hhh - Base Height hundreds of feet
d - Thickness in thousands of feet
(0 indicates to top of clouds)

Turbulence Intensity
0 None
1 Light turbulence
2 Moderate turbulence in clear air, infrequent
3 Moderate turbulence in clear air, frequent
4 Moderate turbulence in cloud, infrequent
5 Moderate turbulence in cloud, frequent
6 Severe turbulence in clear air, infrequent
7 Severe turbulence in clear air, frequent
8 Severe turbulence in cloud, infrequent
9 Severe turbulence in cloud, frequent

Icing Intensity
0 No icing
1 Light icing
2 Light icing in cloud
3 Light icing in precipitation
4 Moderate icing
5 Moderate icing in cloud
6 Moderate icing in precipitation
7 Severe icing
8 Severe icing in cloud
9 Severe icing in precipitation

Fig. B-2. *Military TAFs sometimes included coded forecasts for temperature, turbulence, and icing.*

Index

Illustrations appear in **boldface**.

A

Absolute stability/instability of air, 28–30, **29**
Absorption of heat by Earth, 6
Adiabatic cooling process, 24–27, 80, 116–117
Adiabatic lapse rate, 24, **25**, 61
Advection, 33, 53, 80, 165, 167, 183
Advection fog, 39, **40**
Air mass thunderstorms, 114, 119–120, **120**, **121**
Air masses, 6, 24, 32, 109, 110–114, **112**, 139
Air parcels, 24
Air pollution (*see* haze; smoke; ash/dust/sand)
Air to air visibility, 36, **36**
Air traffic control, 21, 22
Air/fuel mixture, 19–20
Aircraft performance, 13, 17–18
  engine performance vs. temperature, 7, **8**
  temperature vs. 17–18
AIRMET SIERRA, 47
AIRMETs, 37, 98, 99, 170
Airspeed, 13, 14, 17
Airspeed indicator, 13
Alberta clippers, 149
Alert weather watches (AWW), 184
Altimeter settings, 8, 13, 14–17, **15**, 122, 185, 202–203, 206
  density altitude, 17
  errors, 15
  FAR regulations, 14–15
  Flight Level (FL), 16
  Kollsman window, 14
  pressure altitude, 15–16
  Q codes, QNH, QNE, QFE, 16
  temperature vs. performance, 15, 16
  true altitude, 15
  turbulence vs., 17
Altitude, 13, 14–16, 18–20, 74
  lapse rate, 24–27
Alto clouds, 80
Altocumulus castellanus (ACC), 83, 85, **86**
Altocumulus clouds, 81, 83, **84**, 144, 150
Altocumulus floccus clouds, 85
Altostratus clouds, 81, 83, **84**, 145, 150
Anticyclonic circulation, 32, 114, 176, 177, 180–181
Arctic fronts, 148
Area forecasts, 60, 95
Ash/dust/sand, 23, 37, 43–45, **44**, 61, 131, 171, 204
Atmosphere, 1–13, **182**
Atmospheric convection, 33
Atmospheric density (*see* density altitude)
Attitude flying, 74
Automatic terminal information service (ATIS), 22
Average or means in weather, 1
Aviation Routine Weather Report (*see* METAR)
Aviation Selected Special Weather Report (*see* SPECI)

B

Backscatter, 36
Bands of weather, 184
Bjerknes, Vilhelm, weather front theory, 139
Blizzards, 93, 148, 149
Blocking highs, 170
Blue northers, 149
Bora winds, 61
Briefings, 21–22
Broken ceiling, 34
Buell, C.E., 110

C

Calibrated airspeed, 13, 17
Canyon flying, 60, 61, 184
Cap clouds, 62

INDEX

Carbon dioxide, 1
Carburetor heaters, 100–103
Carburetors, icing, 100–103, **101**
Catalina eddy fog, 39
Ceilings, 23, 30, 33–37, 50, 51, 79, 82, 83, 89, 122, 128, 144, 152, 155, 157, 158, 159, 162, 169, 202, 205, 206
Cells, 114
Celsius vs. Fahrenheit temperatures, 7
Chinook winds, 27, 61, 148
Chop, 70
Circulation of air, 53, 54–58, **55**
Cirrocumulus clouds, 81, 85, 86, **88**
Cirroform clouds, 81, 89
Cirrostratus clouds, 81, 85, 86, **87**, 145, 150
Cirrus clouds, 80, 81, 85, 86, **87**, 88, 173
Civil Aeronautics Board, early weather observation, 110
Clear air turbulence (CAT), 70–71, 123, 162, 170
Clear ice, 97
Closed lows, 169, 184
Cloud base ceiling, 34
Clouds, 6, 9, 23, 24, 26, 27, 30, 36, 47–48, **48**, 62, 79–107, 110, 114, 139, 159, 161, 165, 167, 180, 184, 202
  cloud base, 34
  decks between clouds, **51**
Coastal stratus, 39, 49
Cold air masses, 112
Cold fronts, 39, 83, **84**, 120, 140–141, **157**, 183
Cold lows aloft, 169
Condensation, 26, 31
Conditional terms in TAF, 205
Conduction of heat, 6, 80
Confluence, 176
Constant pressure charts, 32, 33, 58, 167
Continental air masses, 111
Contrail (condensation trail), 88, **89**
Convection, 28, 33, 67, 155
Convective LLWS, 65–66
Convective outlooks, 141, 173, 177, 184
Convective SIGMETs, 141, 173, 184
Convective zone, 145
Convergence, 31–32, 109, 110, 113, 114, 122, 163, 173, 176–178, **177**, 179–181, 183
Convergence zone, intertropical, 53, 110
Coordinated universal time (UTC), 200
Coriolis force, 55, 56, 57, 174, 177
Crepuscular rays, 43
Crosswind landings/takeoffs, 58, 72
Cumuliform clouds, 79, 80, 81, 93, 146–147, 152
Cumulonimbus clouds, 81, 88–89, 91, 114, 115, 144, 150, 202, 205
Cumulonimbus mamma clouds (CBMAM), 89, **90**
Cumulus clouds, 43, 63, 67, 80, 81, 88, 89, 115, 145, **146**, 155
Cumulus congestus clouds, 144, **145**, 202
Cut-off lows, 162, 169, 184
Cyclogenesis, 143
Cyclones, 179–181, 183
Cyclonic circulation, 32, 114, 141–143, 146, 176, 179–181, 182, 183

D

Date/time codes, 200
De Coriolis, Gaspard, 55
Density/density altitude, 2, 5, 11–13, 17, 18–20, 53, 139, 170, 195–198, **196–198**
  pressure vs., 12
  moisture vs., 12–13
  temperature vs., 12, 18
Depressions, tropical, 179–181
Dew point, 10–11, **10**, 24–27, 80, 93, 202–203
  adiabatic lapse rate, 24–27
  temperature-dewpoint front (dry line), 32
Diabatic process, 80
Difluence, 176, 177–178
Digging troughs, 173
Direct user access terminals (DUAT), 173
Dirty ridges, 165
Diurnal temperature range, 6–7
Divergence, 31–32, 109, 114, 163, 173, 176–178
Doppler radar (TDWR), 131, 135
Downbursts (*see also* low-level windshear;windshear), 46, 129
Downdrafts, 20, 27, 61, 63, 66, 85, 89, 91, 117, 120, 123, 127, 133
Downslope winds, 58, 61
Downwind landings/takeoffs, 58
Drainage wind, 60
Drizzle (*see also* rain), 45, 92, 148
Droughts, 162, 170
Dry fronts, 145–146
Dry lines, 32
Dry slots, 176
Dust (*see* ash/dust/sand)
Dust Bowl (c. 1930), 162, 170
Dust devils, 44, 45, 111–112, 126

E

Ely Lows, 169
Embedded thunderstorms, 115, 150, 152, 153, 162, 184
En route flight advisory service (EFAS) (*see also* Flight Watch), 21–22
En route weather, 21–22
Engines
  air/fuel mixture, 19–20
  icing, 95, 100–103, **101**
  temperature vs. performance, 7, **8**
Evaporation, 11, 93
Evaporation fog, 40
Evaporative cooling, 68, 69
Exosphere, 3
Eye of hurricane, 180

F

Fall streaks, 86, 92
Fall wind, 61
Ferrel, William, 55
Flight Level (FL), 16
Flight Service Stations (FSS), 21
Flight Watch (*see also* EFAS), 21–22, 73, 74, 159
Foehn wall, 61
Foehn winds, 61, 66
Fog, 6, 30, 34–35, **35**, 37–40, 46, 49, 79, 80, 81, 110, 114, 157, 206
Forecast change groups, TAF, 205–207
Fractocumulus clouds, 82, **83**
Fractostratus clouds, 82, **83**
Free convection, 28, 67
Freezing, 11

Freezing rain/drizzle, 23, 92, 147, 148, 150, 151, 152, 155, 158, 159, 204
Friction force, 55–57, 111, 143
Friction value (mu), runways, 106–107
Frontal turbulence, 70, 74
Frontal zones, 32, 139, 140–141
Frontogenesis, 141, 182
Frontolysis, 143, 182
Fronts, 4, 5, 32, 53, 55, 66, 68, 70, 86, 94, 114, 139, 140–141, 161, 163, 182, 183
 avoidance, 155–159
 cold fronts, 140–141, 143–149, **144**, **157**
 cyclonic circulation, 141–143, 146
 flying through fronts, 155–159
 life cycle, 141–143, **142**
 occluded fronts, 140–141, 142–143, 152–154, **153**, **154**
 overriding, **151**
 passage of a front, 157
 polar fronts, 141, 148
 stationary fronts, 140–141, 154–155
 warm fronts, 140–141, 149–152, **151**, **158**
Frost, 80, 100
Fuel, air/fuel mixture, 19–20
Fuel reserves, 59, 73, 110, 185

G

General circulation of air, 53, 54–58, **55**
Go/no-go decisions, 162
Gravity winds, 60, 61
Greenhouse effect, 11
Ground effect, 18
Ground fog, 37
Groundspeed, 60
Gust fronts, 44, 127, 133
Gustnado, 44, 127
Gusty winds, 58, 61, 64, 74, 122, 146, 148, 157

H

Hadley cells, 57
Hail, 23, 81, 93, 94, 114, 115, 117, 119, 120, 122, 124, 135
Halos around sun/moon, 86, **87**
Hazardous inflight weather advisory service (HIWAS), 21
Haze, 37, 41–43, **42**, 49, 68, 69, 85, 205
Headwinds, 60, 73, 74, 185
Heat as energy, 5
Heat bursts, 27, 128, 155
Heat waves, 162
High pressure areas, 32, 53, 57, 58, 109, 112, 114, 162, 165, 167, 168–172, **171**, **172**, 173
HIWAS, 21
Howard, Luke, cloud classification, 80
Humidity, 10–11, **10**, 26, 32
Hurricanes, 56, 161, 163, 171, 178–181, **178**, **181**, 184–185

I

Ice, 6, 9, 11, 14, 37, 66, 79–80, 85, 86, 88, 89, 93, 95–107, 110, 119, 122, 123, 144, 145, 150–152, 155–158, **157**, 162, 185, 207
 avoidance, 103–107
 carburetor icing, 100–103, **101**
 clear ice, 97
 communications/radios, 96
 conditions producing, 95, 115
 deicing equipment, 98–99
 engine ice, 95, 100–103, **101**
 flying through ice, 103–107
 frost, 100
 hail, 114, 115, 117, 119, 120, 122, 124, 135
 ice protection equipment, 98–99
 induction icing, 100–103, **101**
 instrument icing, 95–96
 mixed ice, 97
 pitot-static system, 95
 propeller ice, 95
 rate of accumulation, 96–98
 rime ice, 97
 runway surface conditions (mu or friction value), 106–107
 severity classifications, 115
 structural icing, 99–100
 supercooled large droplets (SLD), 98
 temperatures, 96
 types of icing, 96–97
Ice fog, 37
IFR flight (*see* ceilings; visibility)
Illusions related to weather, 46–47
Indefinite ceiling, 34
Indicated airspeed, 13, 17
Indicated altitude, 13
Induction icing, 100–103, **101**
Inflight visibility, 37, 36, **36**
Instability of air (*see* stability of air)
Instrument landing systems (ILS), 110
International standard atmosphere (ISA), 7, **8**
Intertropical convergence zone, 53, 110
Inversions, 41, 42, 66, 68, 69, 81, 150
Isobars, 55, 56, 114, 168
Isotachs, 4, **4**
Isotherms, 4, **4**, 165

J

Jet streaks, 172
Jet streams, 3–5, **3**, **4**, 9, 53, 57–58, **57**, 66, 71, 86, 88, 140, 163, 166, 172–173, 177, 184
Jetlets, 172

K

Katabatic winds, 61
Kelvin-Helmholtz (K-H) windshear turbulence, 68–70
Kollsman window/Paul Kollsman, 14
Kraght, Peter E., 109

L

Lake effect precipitation, 6, 79, 95
Laminar flow, 63
Lapse rate, 7, **8**, 24–27, 31–32, 61, 110, 112
Latent heat, 11, 27
Lenticular clouds, 61–64, **62**, **63**, **64**, 83, 86
Level of free convection (LFC), 28, 30, **30**, 67
Lifted condensation level (LCL), 26
Lightning, 114, 115, 117, 122, 124–125, **125**, 134
Limited state thunderstorms, 118–120
Longwave ridges, 163–168, 164–**167**

INDEX

Longwave troughs, 162, 163–168, **164–167**
Low pressure areas (*see also* hurricanes) 32, 53, 57, 109, 112–114, **113**, 141, 143, 162–173, **171**, **172**, 177, 179, 185
Low-level windshear (LLWS), 54, 65–66, 70, 122, 128–133, 184–185, 205
  recovery techniques, 132–133
  takeoff, approach, landing, 132
Low-level windshear alert system (LLWAS), 131

M

Maneuvering speed, 74
Maritime air masses, 111
Maximum/minimum temperatures, 7
Mechanical turbulence, 66–67, 74
Mesoscale convective complexes (MCC), 122, **123**
Mesosphere, 3
METAR, 7, 11, 34, 35, 37, 45, 60, 74, 89, 91, 95, 115, 199–209
Microbursts (*see also* low-level windshear;windshear), 110, 127–129
Minimums (*see* ceilings; visibility)
Mist, 34
Mixed ice, 97
Mode C transponder, 17
Modifiers, report, 200
Moisture, 5, 110, 9–11
  clouds, 9
  cycle of moisture, **12**
  density vs., 12–13
  dew point, 10–11, **10**
  evaporation, 11
  freezing, 11
  greenhouse effect, 11
  humidity, 10–11, **10**
  ice, 9
  latent heat, 11
  precipitation, 9
  relative humidity, 10–11, **10**
  saturation point, 10–11, **10**
  sublimination, 11
  supercooling, 11
  water vapor, 9, 10
Mother of pearl clouds, 88
Mountain waves, 53, 58, **59**, 59, 61–65, 70, 74, 79, 83, 86, 121
Mountain weather, 32, 39, 50, 58, 60–64, 66–67, 94
Mu values, friction values, 106–107

N

Naming hurricanes, 178–179
NEXRAD, 131, 135
Nimbostratus clouds, 80, 82–83, 145, 150
Nimbus clouds, 80
Nor'easter, 149

O

Obscuration phenomena, 184, 202, 208
Obscuration, partial, 34–35, **35**
Obscurement phenomena, 34, 36, 47–48, 50
Obstructions, 201–202, 204, 208
Occluded fronts, 140–141, 142–143, 152–154, **153**, **154**
Omega blocks, 170
Orographic effect, 32
Overcast, 34, 39, 48
Oxygen, 1–2
Ozone, 3

P

Partial obscuration, 34–35, **35**
Performance (*see* aircraft performance)
PIREPS, 21, 66, 74
Pineapple express, 147
Pitot-static system, 13–14, **13**
  icing, 14, 95
Planetary waves, 57
Polar air masses, 111
Polar front, 55, **56**, 110, 148
Polar high, 58
Precipitation, 6, 9, 23, 26, 27, 30, 39, 45–47, 60, 79, 82, 83, 89, 92–95, 115, 117, 122, 134–135, 139, 143, 144, 150–157, 167, 169, 204, 205
Pressure, 1, 2, 5, 8–9, 18, 32, 53, 114
  adiabatic lapse rate, 24–27
  altimeter settings, 8
  density of atmosphere vs., 12
  isobars, 55, 56
  jet streams, 9
  pressure in millibars, 8
Pressure (*Cont.*):
  sea level atmospheric pressure, 8, 9
  standard atmospheric pressure, 9
  temperature vs. pressure gradients, 9
  wind vs. pressure gradients, 9
Pressure altitude, 13, 15–16, 195–198, **196–198**
Pressure gradient force, 55, 57, 61, 145, 177
Prevailing westerlies, 55, 110
Propellers and icing, 95

Q

Q codes, altimeter settings, 16

R

Radiation fog, 37–39
Radiational cooling, 6, 7, 41, 80
Radios and icing, 96
Rain, 6, 23, 34, 37, 39, 45, 60, 83, 85, 89–94, 114–117, 123, 135, 144, 147–150, 153, 162, 165, 170, 178–181, 206
Rain shadow, 94
Rain-induced fog, 39
Ram air pressure, 13–14
Reflection of heat by Earth, 6
Relative humidity, 10–11, **10**, 26–27, 114
Remarks (RMK), 203
Ridges, 32, 53, 86, 109, 114, 162, 163–168, **164–167**, 173, 176, 183
Rime ice, 97
Roll clouds, 91
Rossby waves, 57
Rotor clouds, 61, 63, 65, 70
Runway surface conditions (mu or friction value), 106–107
Ryan, Paul A. Stormscope inventor, 134

S

Sand (*see* ash/dust/sand), 23
Santa Ana winds, 27, 61, 66, 148
Saturation point, 10–11, **10**, 26
Scud, 82, **83**
Sea level atmospheric pressure, 8, 9
Sea/land breeze, 60, 66
See and avoid manuevers, 36

Shear (*see* windshear)
Shelf clouds, 91
Shortwave ridges, 163–168, **164–167**, 182
Shortwave troughs, 162, 163–168, **164–167**, 182
SIGMETs, 44, 45, 98, 99, 141, 170, 173, 184
Sky cover, 202, 204
Slant range visibility, 37, 36, **36**, 49
Smoke, 34, 37, 41–43, **43**, 68, 73, 171
Snow, 6, 23, 34, 37, 45–47, 79, 83, 86, 92–93, 148, 149, 167–168, 170, 204
Soft-field takeoff/landings, 95
Sonoran storms, 181
SPECI, 11
Squalls/squall lines, 58, 89, 121–122, 126, 143–144, 162, 204
Stability of atmosphere, 23, 28–30, **29**, 53, 113
Standard atmosphere, 2, 24
Standard atmospheric pressure, 9
Standard temperature, 7
Standing waves, 61, 62
Static system, 13–14, **13**
Station identifiers, 200
Stationary fronts, 39, 140–141, 154–155, 162, 183
Steady state thunderstorms, 120–122
Steam fog, 40
Stewart, George, author, 178
Stormscope, 134
Stratiform clouds, 30, 79, 80, 150, 152, 154
Stratocumulus clouds, 67, 80–82, **82**, 145, 153
Stratosphere, 3, **3**, 161
Stratus clouds, 49, 80–82, 85, **86**, 143, 146
Stratus fractus clouds, 150, 153
Stream lines, 59
Structural icing, 99–100
Sublimation, 11, 86
Subsynoptic-scale (microscale) circulaton, 53, 58
Supercooled large droplets (SLD), 98
Supercooling, 11
Surface visibility, 36, **36**
Synoptic-scale (mesoscale) circulation, 53, 58

T

Tailwinds, 73, 185
TCU, 89
Temperature, 1, 2, 4, 5–7, 32, 110, 139, 170, 202–203, 207
- absorption of heat by Earth, 6
- adiabatic lapse rate, 24–27
- aircraft performance vs. 17–18
- altimeter settings vs., 15–16
- Celsius vs. Fahrenheit temperatures, 7
- conduction of heat, 6
- density altitude, 18–20
- density of atmosphere vs., 12
- diurnal temperature range, 6–7
- engine performance vs., 7, **8**
- fog formation, 40
- heat as energy, 5
- heat burst, 27
- international standard atmosphere (ISA), 7, **8**
- isotherms, 4, **4**
- lapse rate, 7, **8**, 24–27
- latent heat, 11
- maximum/minimum temperatures, 7
- pressure vs., 9
- radiational cooling, 6, 7
- reflection of heat by Earth, 6
- standard temperature, 7
- terrestrial radiation, 6, 7
- warm/cold are relative terms, 5–6

Terminal aerodrome forecasts (TAF), 7, 37, 60, 95, 199–209
Terminal Doppler weather radar (TDWR), 131, 135
Terrain obscurement (*see* obscurement phenomena)
Terrestrial radiation, 6, 7
Texas norther, 149
Thermal advection, 165
Thermal lows, 114, 171
Thermal turbulence, 67–68, **68**
Thermosphere, 3
Thunderstorms, 54, 58, 59, 60, 65, 83, 89–94, 109–110, 113–128, 140–162, 167, 169, 177, 180, 184, 204–206
- air mass thunderstorms, 119–120, **120**
- altimeter settings, 122
- avoidance, 133–137
- ceilings, 122, 128

Thunderstorms (*Cont.*):
- cells, 114
- clear air turbulence (CAT), 123
- conditions producing, 115
- downbursts, 129
- downdrafts, 117, 120, 123, 127, 133
- embedded thunderstorms, 115
- flying thunderstorms, 133–137
- gusts, gust fronts, 122, 127, 133
- hail, 114, 115, 117, 119, 120, 122, 124, 135
- hazards of, 122–128
- heat bursts, 128
- ice, 119, 122, 123
- life cycle, 115–118, **116**
- lightning, 114, 115, 117, 122, 124–125, **125**, 134
- limited state thunderstorms, 118–120
- low-level windshear (LLWS), 122, 128–133
- mesoscale convective complexes (MCC), 122, **123**
- microbursts, 127, 128
- nighttime storms, 122, 128
- NTSB guidelines, 99, 135
- precipitation, 115, 117, 122, 134–135
- rain, 114, 115, 117, 123, 135
- severity classifications, 115
- squall lines, 121–122, 126
- steady state thunderstorms, 120–122
- storm detection equipment, 134
- Stormscope, 134
- tornadoes, 120, 122
- turbulence, 119, 122–123, 128
- updrafts, 116, 117, 120, 123, 133
- VIP levels, 134
- visibility, 122, 123, 128
- winds, 114, 115
- windshear, 126–133

Time/date codes, 200
Tornadoes, 120, 122, 131, 149, 177, 205
Torricelli, Evangelista, inventor of barometer, 8–9
Towering cumulus clouds, 89, **90**, 143–144, **145**, 150, 202
Transcribed weather broadcast (TWEB), 141
Transponders, 17
Tropical air masses, 111

Tropical cyclones, 179–181
Tropical depressions, 179–181
Tropical storms, 179–181, 185
Tropopause, 3, **3**
Troposphere, 2, **3**, 161, 179
Troughs, 32, 53, 70, 109, 113–114, **113**, 139, 161, 162, 163–168, **164–167**, 173, 176, 182–183, 184
True airspeed, 13, 17
True altitude, 13, 15
Tule fog, 37–39, **38**
Turbocharging, 19
Turbulence, 3, 5, 17, 20, 40, 54, 58, 61–71, 74, 81–91, 114, 119, 122–123, 128, 144–146, 155, 156, 161, 162, 184–185, 207
- clear air turbulence, 70–71
- degrees of, 65
- frontal turbulence, 70, 74
- locations/conditions favorable to, **75–76**
- mechanical turbulence, 66–67, 74
- mountain waves, 74
- thermal turbulence, 67–68, **68**
- wake turbulence, 71, **72**, 76–77
- windshear turbulence, 68–70, 71, 74

U

Ultraviolet radiation, 3
Updrafts, 61, 63, 81, 82, 89, 91, 93, 116, 117, 120, 123, 133
Upper-level highs, 168–172, **171**, **172**
Upper-level lows, 168–172, **171**, **172**
Upper-level ridges, 163–168, **164–167**
Upper-level troughs, 163–168, **164–167**
Upper-level weather systems, 161–185
Upslope fog, 39, **41**
Upslope winds, 94

V

Valley winds, 60, 66
Venturi effect, 59
Vertical motion in atmosphere, 31–33
Vertical speed indicator, 13
VFR flight (*see* ceilings; visibility)
VIP levels, 134
Virga, 83–85, **85**, 92, 130
Visibility, 23, 30, 33–37, 48–52, 81–83, 89, 91, 114, 122, 123, 128, 144, 151–155, 157–159, 162, 169, 201–206
Volcanic ash (*see* ash/dust/sand)
Volcanic Ash Forecast Transport and Dispersion Chart, 45
Vortex, 71, **72**
Vorticity, 32, 56, 57, 163, 173–176, **174**, 184

W

Wake turbulence, 54, 71, **72**, 76–77
Wall clouds, 91, 180, 184–185
Warm air masses, 112
Warm fronts, 39, 140–141, 149–152, **151**, **158**, 183
Water, 1, 9, 11
Water bodies, 6, 40, 58, 95
Water vapor, 1, 6, 9, 10, 11
Waterspouts, 112, 126
Westerlies (*see* prevailing westerlies)
Whirlwinds, 45, 111
Whiteout, 46–47, 93
Winds (*see also* hurricanes), 2–5, 4, **9, 20, 27, 32, 39, 43–45, 53–77, 81, 114, 115, 139, 144, 145, 150–154, 159, 162, 173, 176, 184, 185, 200–201, 204, 206**
Winds aloft, 57, 58, 59, 60, 159, 176, 182
Windshear, 5, 44, 53, 54, 58, 61, 65–71, 74, 85, 91, 110, 126–133, 141, 184, 204, 205

Z

Zero-zero conditions, 38

About the Author

Terry T. Lankford has been a certified pilot for 30 years. He is a full-time flight instructor, and possessor of the coveted Gold Seal. Lankford is multi-engine certified and has owned and operated an FAA-approved ground school in Los Angeles. He also has owned a Cessna 150 and has flown across the country and in Hawaii, Canada, Mexico, and Europe. In addition, he is an air traffic control specialist and a flight service station specialist and has studied weather at the FAA Academy.